T0176446

FREE SPACE OPTICAL NETWORKS FOR ULTRA-BROAD BAND SERVICES

FREE SPACE OPTICAL NETWORKS FOR ULTRA-BROAD BAND SERVICES

Stamatios V. Kartalopoulos

A JOHN WILEY & SONS, INC., PUBLICATION

Published by John Wiley & Sons, Inc., Hoboken, New Jersey.
Published simultaneously in Canada.

For general information on our other products and services or for technical support, please contact our Customer Care Department within the United States at (800) 762-2974, outside the United States at (317) 572-3993 or fax (317) 572-4002.

Wiley also publishes its books in a variety of electronic formats. Some content that appears in print may not be available in electronic formats. For more information about Wiley products, visit our web site at www.wiley.com.

Library of Congress Cataloging-in-Publication Data is available.

ISBN: 978-0-470-64775-2

oBook ISBN: 978-1-118-10423-1
ePDF ISBN: 978-1-118-10421-7
ePub ISBN: 978-1-118-10422-4

Printed in the United States of America.

10 9 8 7 6 5 4 3 2 1

To my wife Anita for her love and support

CONTENTS

PREFACE

There has been a need for transmitting messages since the beginning of human history. However, the speed of communication—that is, sending a message from point A to a remote point B reliably and securely—has always been extremely important. It was soon recognized that the fastest means to send a message was not the fastest horse or the fastest runner, but light. Thus, in some ancient countries such as Greece, a network of light-towers was established on top of hills and mountains; by signaling from tower to tower using lit torches, an encoded message could be transmitted effectively "at the speed of light". Today, this network has been dubbed the *Agamemnon's Link* as it makes reference to the communication practices during the Trojan era [Aeschylus, Polybius, etc.].

Advancements in technology have produced lasers, tiny semiconductor devices, which generate a very narrow beam of light in the visible (400–700 nm) or in the invisible spectrum (800–900 nm and 1280–1620 nm); the simple laser pointer is just one of the many applications of lasers. One of the most important applications of lasers has been optical communication networks.

The fiber optic communications network has enabled an unprecedented amount of information to flow, unleashing the imagination and creativity of both technologists and communications service providers. Today, what was considered science fiction a few decades ago has become reality; that is, multi-play and 3-D telecommunication services. Similarly, while today's telecommunication services will be considered archaic in a decade or so, many futuristic and fictional services will be reality.

The unprecedented amount of information flow that the optical backbone communications network made possible created a problem at the access network, dubbed the "first/last mile" bottleneck; the access network was not retrofitted with fiber as fast as the backbone, and the copper-based loop plant that was designed for analog signals was not suitable for ultra high data rates. This bottleneck was foreseen back in the 1980s, when research and prototyping proved that existing analog loops could be used for very high digital data rates, if properly conditioned; this research and proof of concept was demonstrated by AT&T Bell Labs scientists, where I was privileged to be on the epicenter of this activity. The fruits of this effort were what is now known as integrated services digital network (ISDN) and also as digital subscriber loop (DSL), which in the period 1990–2010 addressed the "first/last mile" bottleneck.

During these last two decades, the initially simplistic Internet was becoming more popular, and new more potent data systems (routers) were interconnected with fiber to create a data network layer. In addition, new protocols were designed allowing data rates in the Gigabit per second region, and new generation cellular wireless protocols broke the telephony boundary and offered data services and image over electro-magnetic waves. As a consequence, the deliverable data rate to end users increased, the traffic demand by end users increased, and the types of services that were offered to end users became more complex. As a result, the access network required much higher digital data rates in order to meet the demand and type of services, which only optical technology could offer. Currently, the fiber to the premises (FTTP) and passive optical networks (PON) are promising to address this in both residential and enterprise applications.

Thus, optical solutions for ultra-high data rate deliverability are based on fiber. Yet, there are applications for which fiber may not be a viable solution, either because the fiber infrastructure is not in place, or because the anomalous terrain is not amenable to fiber installation, or because a high data rate network is needed for a short period that does not justify the fiber infrastructure investment. In such cases, there are two solutions: RF direct links that offer few Mb/s over relatively short distance, or RF satellite links that offer few Mb/s over longer distance but via a satellite. However, RF solutions depend on available spectrum, which is getting saturated by new wireless services.

A different solution is the free space optical (fiber-less) or FSO technology, also called "open-air optical communications" and "optical wireless"; FSO capitalizes on existing technology and components (lasers, receivers and optics) and well-established protocols (Ethernet, SONET/SDH, ATM) to quickly establish optical links that support Gb/s over a distance of few kilometers in direct links with line of sight, or many kilo-meters in concatenated links. As such, the FSO technology can be used to quickly bring ultra-high data rates to the residential and to the enterprise domain.

Strictly speaking, the term FSO implies optical communications well above the troposphere and atmosphere (hence *free space*); in fact, originally this technology was developed for NASA in inter-satellite applications that envisioned satellites inter-linked to a network in the sky via inter-satellite modulated laser beams, a concept that I also had the privilege to develop at Bell Labs. However, the idea of linking communication stations via laser beams was also attractive to military terrestrial applications from where they migrated to civilian applications. But in this case, the Earth's atmosphere and particularly the moody troposphere imposes to the laser beam several impairments, such as attenuation and more, and thus the term "open-air optical communications" may describe FSO better; however, for traditional purposes in this book we adopt the term FSO in a more generic sense.

But what is FSO? Well, think that you have a laser-pointer that aims at a specific point in space where a photodetector is placed. The photodetector senses the light of the laser-pointer and it produces a constant electrical output. This is a laser link but as described is *information-less*. Now, think that you turn the laser on and off to modulate the beam, as in Morse-code or ASCII code. The photodetector now produces an electri-cal output that is modulated as the laser beam is. Now, this link carries information. Think further that the beam is modulated very fast, millions of times or billions of times per second. If pulses in the beam have enough energy to reliably produce electrical pulses at the output of the photodetector, you have just built an impressive FSO communications link at Mb/s or Gb/s data rate. Now, to go one step further, imagine that you have several laser-pointers, each a different color and each modulated independently to carry different information. Think also that all these different colored beams are multiplexed into a single composite beam and transmitted. At the destination, the composite beam is demultiplexed to its color components and each component is directed to a separate photodetector. What you have just built is the next generation FSO link known as wavelength division multiplexed FSO (WDM-FSO).

How easy is it to build? Obviously, the aforementioned example has trivialized FSO technology in an attempt to explain the basic principles of operation. In reality, there are many issues to address before one can claim that this is a commercial grade FSO system. For example, if the distance between the laser and the photodetector is a few kilometers long, then how can the beam be directed onto the target photodetector, which may be less than a centimeter square, with high precision? Conversely, if the laser beam is slightly divergent so that after a few kilometers the beam is few meters in diameter wide, then how much energy impinges onto the tiny target photodetector? Will the impinging energy be enough to produce electrical pulses reliably? There are many more issues that one needs to address in order to establish a reliable commercial grade FSO link.

Furthermore, if FSO is integrated with secure wireless technology and protocols, such as WiMAX or Wi-Fi or cellular technology, then it can also offer end user mobility. Thus, with wireless technology, at minimum three highly desirable features are supported: quick deployment, ultra-high data rate to end user, and end user mobility.

In this book, we examine the technology that goes into a FSO link, a WDM-FSO link, and various associated issues. We examine issues associated with atmospheric phenomena that affect the laser beam propagation. We examine practically useful FSO network topologies, such as point to point, ring and mesh, and issues associated with each topology. We examine the design complexity of FSO nodes, traffic capacity and fault protection and types of service that FSO technology can support. We also examine security issues associated with FSO technology and networks.

In this book, we also describe the strengths and weaknesses associated with FSO technology, FSO components necessary for the design of FSO nodes, FSO protection strategies and network reliability, and FSO integration with wireless and cellular technologies for service portability, for ad-hoc networks and for end user mobility.

FSO is still evolving and therefore we hope that this book will spark the interest of the reader, stimulate thinking, and invoke many questions that will bring FSO technology and networks to the next level of network intelligence and complexity or simplicity. Happy reading. . . .

STAMATIOS KARTALOPOULOS

ACKNOWLEDGMENTS

To my wife Anita, son Bill and daughter Stephanie for consistent patience and encouragement. To my publishers and staff for cooperation, enthusiasm, and project management. To the anonymous reviewers for useful comments and constructive criticism. And to all those who worked diligently on the production of this book.

ABOUT THE AUTHOR

Stamatios V. Kartalopoulos, PhD, is the Williams Professor in Telecommunications Networking with the University of Oklahoma, EE/ECE/TCOM graduate program. His research emphasis is on optical communication networks (including access PON/FTTH and FSO), optical technology and optical metamaterials, optical network security, including quantum cryptography and quantum key distribution protocols, chaotic processes, chaotic quantum networks, and biometric networks.

Prior to academia, he was with Bell Laboratories where he defined, led and managed research and development teams in the areas of DWDM networks, SONET/SDH and ATM, Cross-connects, Switching, Transmission and Access systems, for which he received the President's Award and many awards of Excellence.

He holds nineteen patents in communications networks, has authored nine reference textbooks and more than two hundred scientific papers, and he has contributed chapters to many books; his latest book on Network Security received the *2009 Choice Award of Outstanding Academic Titles,* and his 2005 paper on PON networks received the *Award of Top Cited Article in the period 2005–2010*.

He has been an IEEE and a Lucent Technologies Distinguished Lecturer, and has lectured worldwide at universities, NASA, conferences and enterprise entities. He has been keynote and plenary speaker at major international conferences, has moderated executive forums, has been a panelist of interdisciplinary panels, and has organized symposia, workshops and sessions at major international communications conferences.

Dr. Kartalopoulos is an IEEE Fellow, founder and past chair of the IEEE ComSoc Communications & Information Security Technical Committee (CIS TC), past chair of ComSoc SPCE and of Emerging Technologies Technical Committees, member at large of IEEE Biometrics Council, series editor and past editor-in-chief of IEEE Press, Area-editor of IEEE Communications Magazine/Optical Communications, member of IEEE PSPB, VP of IEEE Computational Intelligence Society, and member of the IEEE Experts-now program.

INTRODUCTION

I.1 IN PERSPECTIVE

Optical communication is a technology capable of transporting an unprecedented amount of information per second in an optically transparent medium at "the speed of light". In a simple form, the use of light as an information transportation vehicle has been used for millennia, with the first written evidence making reference to the epoch of the Trojan War (c.1200 BCE) by Aeschylus in his theatrical play "*Agamemnon*":

> And I am waiting for the signal torch,
> That flame of fire.
> From Troy bringing the news,
> The tidings of its fall . . .

Free Space Optical Networks for Ultra-Broad Band Services, First Edition. Stamatios V. Kartalopoulos.
© 2011 Institute of Electrical and Electronics Engineers. Published 2011 by John Wiley & Sons, Inc.

Today, this optical communications link is dubbed *"the Agamemnon's Link"*. Optical communication links were also used with similar methods and added security features using time-dependent encryption codes at each network node (actually, these nodes were optical towers on top of hills or mountains). One method that was developed by the military scientist *Aeneias Tacitos of Stymphalos* (4th c. BCE) was based on the "leaky bucket", or *clepsydra*, principle; each node would change the encryption key according to the water level in a leaky ceramic jar. Another method used several torches at each node to encode the optical signal, a method that was invented by *Cleoxenos* and *Demokleitos*, and it was perfected by Polybius (203–120 BCE); today, this method is known as the *Polybius Square* [1].

Similarly, when an optical communications link was needed over a body of water, such as from land to ship and/or ship to ship, the encoded message was transported using either the sunrays or a torch and a polished shield (ancient Greek shields were made of bronze, many of which were polished).

The aforementioned ancient optical communication technologies used light that propagated in the atmosphere. Today, we use laser devices to generate a narrow beam of almost monochromatic light, photodetectors to detect light, other optical and electronic components in order to communicate in the atmosphere, and especially the troposphere; we call this technology Free Space Optical (FSO) to distinguish from another optical communications technology that uses similar or same devices but in this case light propagates in dielectric fiber [2, 3].

I.2   FIBER NETWORKS AND THE LAST/FIRST MILE BOTTLENECK

Today, the accelerated progress of microelectronics and electro-optical devices has made yesterday's science fiction today's reality. Although the speed of light, in terms of data propagation in a medium, has not changed the amount of data, the unit of time per bit has changed by an astonishing factor that results in many zeroes next to one, from 1,000,000 to 1,000,000,000, or from Megabits per second (Mb/s) to Gigabits per second (Gb/s). To put this in perspective, transmitting data at 1 Gb/s means that 1 Gigabit file is transmitted in just one second, or equivalently, the full contents of 24 volumes of a typical encyclopedia in just one second. This great achievement is the result of advancements in laser devices, ultra-fast modulators, ultra-sensitive photodetectors, all-optical amplifiers, optical switching, ultra-fast switching electronics, ultra-pure fiber, and many more. In just two decades these advancements are responsible for incredible developments in circuit integration, optical transceivers, and optical systems and networks capable of transmitting huge amounts of data in a second, meeting the stringiest expected performance-cost. As a consequence, in about a decade, optical communications has become the only technology of choice in the backbone network (fiber has replaced copper 100%). However, one cannot support the same argument for the access network, and for reasons of cost (capital and maintenance) versus profit. As a consequence, the amount of potential data that could be transported over the backbone network was by several orders of magnitude the amount of data that could be delivered to all end users, thus creating the so called "last/first mile bottleneck".

Within just a few years, the Internet has been exploding, new services over the Internet have been offered, traditional telecommunication networks support new integrated services, and wireless digital technology, mobile and semi-mobile, has added new services that have aggravated the bottleneck.

To overcome the bottleneck and deliver more digital data to end users, two major efforts have been made; the Digital Subscriber Loop (DSL) based on existing telephony copper loops and the Fiber to the Premises (FTTP) based on modern optical communications network [4, 5].

- DSL is a digital transmission technology and it has been deployed for the last two decades. DSL uses existing loop copper twisted pairs (TP) and depending on distance from the central office (CO) to the end user, delivers digital data up to few Mb/s.

- FTTP is an optical transmission technology that uses single mode fiber (SMF) from the CO to the neighborhood optical line termination (OLT), and from there using SMF or multimode fiber (MMF) to the optical network termination (ONT) at the end user premises. FTTP may be constructed with passive optical components, in which case it is known as passive optical network (PON), or with passive and active optical components. FTTP is superior to DSL in terms of data rate and distance, and it is capable of meeting current and future customer demand of existing and new services that require ultra-high data rates, or Ultra-Broad-Band (UBB). Such services are high-definition two-way interactive video, ultra-high speed data (including fast Internet with interactive video, voice and data), as well as the traditional services of voice and sound (music).

I.3 ALTERNATE ACCESS TECHNOLOGIES

Currently, the fiber optic technology has successfully met all expectations and the prediction is that it will not be replaced by another technology for many years to come. However, there are certain applications that require immediate ultra-high data rate services, but either there is no existing fiber deployed, or there is insufficient fiber infrastructure. In this case, and until fiber is deployed, another optical communications technology is needed rapidly. Such technology is the Free Space Optical (FSO), Figure I.1.

The FSO is an outdoor wireless communications technology that promises data rates higher than 1 Gb/s per link, and if wavelength division multiplexing (WDM) is incorporated, the aggregate bandwidth per link may exceed 10 Gb/s and potentially 50 or 120 Gb/s per link. In most cases, a link is established from rooftop-to-rooftop of tall buildings, from window-to-rooftop or from window-to-window, and in some cases FSO nodes are mounted on tall poles. Although these ultra-high data rates may seem very high for residential applications, there are future applications that potentially may demand even higher bandwidth; one such application is three-dimensional high definition video.

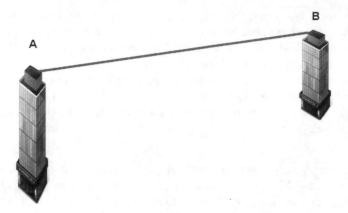

Figure I.1. The point-to-point FSO link concept.

FSO technology uses a modulated laser beam as the transmitter and a sensitive photodetector as the receiver. A simple communications link is constructed with a laser beam directed to an (unobstructed with line of sight (LoS)) photodetector; two laser-photodetector transceivers comprise a two-way, full-duplex FSO link. Because the laser beam has a wavelength in the micrometer range, and not in the traditional electromagnetic radio spectrum (meter-millimeter), there is no need for FCC or municipal license approvals. As a consequence, FSO links and FSO networks can be set up and be deployed rapidly, in a matter of few days as compared with years for a new fiber network. The FSO cost is also very low as compared with cost of fiber, if one considers the time and cost to obtain right of way permits (private, municipal, enterprise), and the labor intensive fiber planning and deployment.

Current FSO links operate at data rates that range from Mb/s to Gb/s, and possibly 10 Gb/s over a link length that ranges from few hundred meters to few kilometers. Typical transmission signals are based on a standard and popular protocol such as SONET/SDH, T1 (1.544 Mb/s, or E1 2.048 Mb/s), E3/DS3, and 10/100/1000 or higher Ethernet. These rates are popular in applications such as:

- Last/First mile connectivity to the access network
- LAN to LAN inter-connectivity (1GbE), for campus or Metropolitan applications.
- Mobile station to network connectivity
- Emergency communications network deployment
- Disaster relief network application
- Expedient and semi-permanent network deployment
- Inter-satellite communication
- Satellite to/from earth station communication
- Mobile to stationary station communication
- Deep space communications [6, 7]

Another possible application is the use of platforms positioned at high attitude (about 20–30 km); such platforms can communicate via FSO links with deep space platforms because above them there are no clouds or troposphere to interfere with laser light.

However, the laser beam propagates through the atmosphere, and in most cases through its lower layer known as troposphere, which is not a well-defined stable medium like the dielectric fiber. The troposphere is not a stable medium over time and thus its physics, chemistry and varying parameters should be well-understood and considered in FSO link design to warrant a link operation at the expected performance and efficiency.

Although FSO requires LoS, the radio technology does not. The reason is that RF electro-magnetic waves propagate through walls and around corners. Because of this, radio waves have been, from the outset, applied in mobile communications whereby a moving transceiver may communicate with another moving or with a stationary transceiver. Thus, mobility and freedom of wires has been the major advantage of wireless mobile communications technology. However, its major disadvantage is the low product {*bit rate*} × {*distance*}, where the bit rate is several Kb/s and the distance few kilometers at its best; the higher the bit rate the shorter the distance between end user and the stationary antenna. Comparing this with the product of FSO, FSO is orders of magnitude better. In FSO, the distance is many kilometers and the bit rate is many Mb/s.

However, FSO and wireless mobile technology are complimentary: the FSO delivers ultra-high bandwidth to the end user and it lacks user mobility, whereas the wireless delivers low bandwidth to the end user and it allows for mobility within a range, and also unlimited mobility if efficient and secure handoff protocols are used, Table I.1.

As an example, consider integrating FSO with WiMAX; this is a logical step that takes advantage of the best of both technologies if a fiber network is not already present. If fiber is present, then FTTP with WiMAX may be the answer to deliver bandwidth many times 10 Gb/s at a link length without repeaters up to approximately 30 Km. In FSO with WiMAX, FSO is used to reach the access domain (neighborhood, multi-level building) and from there, the wireless technology to deliver high data rates to thousands of mobile users within a range of few kilometers, Figure I.1. Moreover, because the integration of FSO with the fiber-optic backbone network uses similar technologies, both technologies are within the vision of the next generation all-optical modern network with standard and popular protocols.

So, how does a FSO transceiver work? How do atmospheric phenomena impact the propagation of light? What kind of "light" is used? How safe is it? And, what

TABLE I.1. FSO vs Wireless Mobile

	Bandwidth	Link Length	Mobility
FSO	High	Long	None
Wireless Mobile	Low	Short	Limited to High

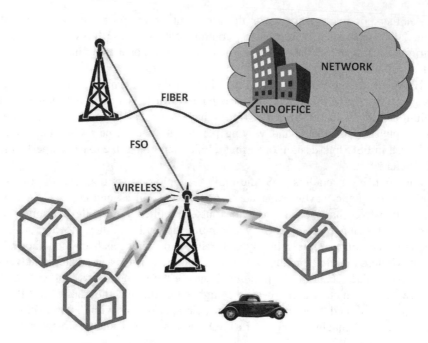

Figure I.2. FSO with Wireless end user connectivity.

happens if there are severe tropospheric phenomena? How can we construct a network, and what type of network, with FSO transceivers? We take a brief look and try to provide a quick answer, yet examine each question in detail in the chapters that follow.

I.4 FSO WAVELENGTH

Currently available Free Space Optics (FSO) transceiver systems establish a two way communication over a link. Depending on the wavelength they use, FSO systems fall into two categories: systems that operate within the 800–900 nm range, and those that operate at 1310 or near 1550 nm. The first uses VCSEL laser technology that is low cost and low power and thus more suitable for low rate (Mb/s) and short distances. The second uses improved vertical cavity surface emitting lasers (VCSEL) or distributed feedback (DBF) lasers that emit at higher optical power (typically at −8 dBm), yet within the power limits and standard wavelengths that are defined for communications (assuring compatibility and interoperability), for eye safety, and for reduced solar background radiation. FSO laser beams use standard telecommunication lasers at low power and thus there is no safety issue associated with it.

The photodetector at the receiver is typically an avalanche photodetector (APD) with sensitivity as low as −40 dBm. From fiber-optic communications, it is well-known that the sensitivity of the APD detector greatly depends on the impinging wavelength and also on the bit rate. For example: at 1550 nm, the APD sensitivity for acceptable performance at 2.5 Gb/s is −29 dBm, at 1 GbE (actual 1.25 Gb/s) is −33 dBm, and at OC-3 (155 Mb/s) it is −43 dBm.

With respect to interference with solar background radiation (SBR) in FSO communications, it should be pointed out that SBR becomes a serious issue when the transceiver faces directly east or west and where the sun emerges or sinks the horizon; SBR can be about ~6000°K. Thus, in many applications this may not be an issue. In addition, if 1550 nm is used in FSO, then a long-pass optical filter can reject most wavelengths below 1300 nm; the visible spectrum is approximately within the band 400 (violet)–700 (red) nm, and UV is smaller than 400 nm. That is, SBR may be addressed with both smart enclosure and filter design and with transceiver positioning; few degrees off-axis from the solar direction greatly reduces SBR.

In the simplest configuration, think of a laser device that generates a thin beam of light in the wavelength band 1300 to 1600 nm, and modulated at a data rate of 1 Gb/s. This beam is directed to a photodector that is located from few hundred meters to 2 kilometers away. When the beam arrives at the photodetector, two major key points are observed. First, the diameter of the cross section of the beam at the receiver site is much larger than the diameter of the beam at the transmitter site as a result of beam spatial divergence; that is, the originally thin beam, less than a millimeter thin, is now more than a meter or two wide causing the optical power per square unit to be reduced as the square of the distance from the laser. Second, the photodetector itself has a very small area, several millimeters square, as compared to several meters square cross section of the beam at the photodetector site; that is, the amount of power that impinges on the photodctector is a fraction of that arriving and therefore the photodetector may not be sensitive enough to respond to few photons impinging on it. As a result, a wide aperture telescope may be used to focus more photons onto the photodetector.

In addition to this, the FSO laser beam at wavelength above 1400 nm is absorbed by the lens and by the cornea of the human eye, and thus there is no destructive focal point to create damage on the retina. As a consequence, lasers emitting at above 1400 nm may be as much as 50 times more powerful than lasers below it. Thus, lasers at 1550 nm (a popular wavelength in optical telecommunications) can be 50 times stronger than lasers at 800 nm and exhibit the same eye safety. This also implies a 17 dB (10log50 db) margin benefit of 1550 nm as compared to 800 nm, which is beneficial in link length engineering, link bandwidth and link power budgeting. Moreover, 1550 nm penetrates glass panels more efficiently (less insertion loss) making it more suitable for window-to-window or window-to-rooftop link establishment; in specific applications, it is more favorable to deploy window-to-window links if line of sight exists.

I.5 TRANSMITTER-RECEIVER LINK

FSO systems should be designed to operate under various atmospheric conditions, such as attenuation, fog, haze, rain, snow and temperature variability. In general, longer wavelengths (1550 nm) are more advantageous than shorter wavelength (800 nm) in haze or light fog condition; in very low visibility (heavy fog or dense haze), this advantage does not apply. However, the 17db margin of 1550nm systems has deeper penetration in fog or other atmospheric attenuation mechanisms.

In addition to the above margin, which is a benefit of the wavelength in use, the optical detector is equally important. In this case, the type of photodetector and the optics associated with it play a synergistic role for efficient detection. FSO detectors may be P-intrinsic-N (PIN) or avalanche photodetecor (APD) devices.

- The PIN photodiode consists of an intrinsic (lightly doped) region sandwiched between a p-type and a n-type doped semiconductor material. When it is reversed biased, its internal impedance is almost infinite (as an open circuit) and its output current is proportional to the input optical power. When a photon enters the intrinsic region, it creates an electron-hole pair, and the PIN produces a current pulse with duration and shape that depends on the R-C time constant of the PIN device. The capacitance of the reversed biased PIN photodiode is a limiting factor to its response (and switching speed). At low bit rates (<Gb/s), the parasitic inductance of the PIN may be neglected. However, as the bit rate becomes higher, parasitic inductance becomes significant and causes "shot noise".
- The avalanche photodiode (APD) is a semiconductor device that consists of a two-layer semiconductor sandwich where the upper layer is n-doped and the lower heavily p-doped and in operation it is equivalent to a photomultiplier. At the junction, the charge migration (electrons from the n and holes from the p) creates a depletion region and from the distribution of charges a field is created in the direction of the p-layer. When the APD is reverse biased and no light impinges on the device, then, due to thermal generation of electrons, a current is produced known as "dark current," which is manifested as noise. If the reversed bias device is exposed to light, then photons reach the p-layer and cause electron-hole pairs. However, because of the strong field in the APD junction, the pair flows through the junction in an accelerated mode. The electrons gain enough energy to cause secondary electron-hole pairs, which in turn cause more. Thus, an *avalanche* process takes place (hence its name) and a substantial current is generated from just few initial photons. As a consequence, APD devices are about four times more sensitive than PIN devices, although PINs are lower-cost devices.

The associated optics with the receiver is also important. A larger aperture receive lens collects more photons that are focused onto the photodetector, as well as reduces errors by averaging due to atmospheric turbulence as a result of solar loading and natural convection that cause dynamic micro-temperature variations, and thus temporally and spatially varying refractive index changes in the air on the propagating path of the beam; this is known as *scintillation*. Scintillation is similar to the apparent twinkling of distant lights, also known as *stilbe*, which in FSO propagation is manifested by error bursts due to dynamic variation of the signal power at the receiver and thus by a dynamic variation in signal to noise ratio (SNR) that increases bit error rate (BER). A large aperture of the receiver lens collects more photons over a larger beam area and it dynamically averages the signal power. Some manufacturers of FSO transmitters encompass multiple lasers to create multiple parallel beams for higher power, better protection to laser failure and minimized scintillation contributing effects.

Yet, having an excellent transmitter and an excellent receiver is still not enough. The link can be efficient if the transmitted beam continuously aims the photodetector like a cross-hair on a target; that is, the alignment of the transmitter-receiver link is extremely important, as well as the center of the beam to impinge the receiving photodetector all the time even if the photodetector or the transmitter sways by few meters due to strong wind (very tall buildings may sway by a couple of meters at the rooftop); the latter implies that an automatic self-tracking mechanism should exist that maintains a continuously perfect alignment of the transmitter-receiver pair. Clearly, if self-tracking is not employed, sway by a meter or so will destroy the alignment of the transmitter-receiver pair and the link will become inoperable.

I.6   THE ATMOSPHERE

The atmosphere, and particular the troposphere, is the medium in which the FSO laser beam propagates. This medium is a dynamic medium that continuously changes in chemical composition, temperature, humidity, pressure and air movement over time and along the path of the propagating beam. Rain, fog, haze, smog, snow, and other conditions have an adverse effect on the beam and thus on the communication signal.

For example, extreme temperature conditions may disrupt the FSO link operation. The outdoor housing of the FSO transceiver is affected by low and by high temperatures and also by possible accumulated frost; typical operation the operating temperature of the opto-electronics is −30°C to 60°C (with built-in defroster).

- It is possible for frost to accumule on the housing due to humidity and low temperature. In this case, an appropriately designed, water-sealed and shelf-controlled heated housing is recommended. In extremely high temperatures, a self-cooled housing is required to maintain the opto-electronics operating within the manufacturer operating temperature limits. An automatically temperature controlled housing keeps the opto-electronics operating with the operating and humidity range and it extends the lifespan of the components.

- Dense fog and heavy snow may disrupt the FSO link operation. A communication link down for extended period is unacceptable. Automatic RF back-up is a protection strategy that keeps the link operable, although downgraded, for as long as an adverse atmospheric condition persists. However, in this case, because the millimeter-wave RF beam may be 20 or more meters in diameter, as compared to about two meters of the substituted optical beam, the communication link security needs to be considered.

- Link engineering considers transmitted power, propagating beam-width, receiver sensitivity, receiver optics, and alignment strategy in such a balanced way as to minimize the mean down time due to atmospheric conditions. Link engineering also considers the mean lifetime of components before failure, or the mean time before failure (MTBF) so that a prediction may be made of link failure due to degraded components. The MTBF data for each component is typically provided by the manufacturer.

Overall, FSO is an optical communications technology with the advantage of, compared to fiber optic, quick deployment, standard optical telecommunications components utilization and low operating cost and maintenance. However, it has the disadvantage of been susceptible to atmospheric conditions, and it requires a housing design that maintains the transmitter operating within the recommended by the manufacturer limits, temperature, humidity and power. Moreover, it requires superb alignment and self tracking to maintain continuous line of sight.

Similarly, FSO, compared to mobile wireless technology, delivers ultra-high bandwidth to stationary end users and over relatively long links, whereas mobile wireless delivers much lower bandwidth to mobile end users and over long distances if handover protocols are used, or limited distance without handover.

As a measure of goodness of the FSO technology with other access technologies (from the user perspective), if the normalized cost/DataRate-month for the dial up technology is 1.00 unit, then for cable modem it is about 0.10 unit, and for FSO is about 0.01 unit (we do not have sufficient data for FTTP).

In the chapters that follow, we examine the FSO components, the medium, network, topologies, traffic, protection, applications, engineering, and security issues in more detail.

REFERENCES

1. S.V. Kartalopoulos, *Security of Information and Communication Networks*, IEEE/Wiley, 2009.
2. S.V. Kartalopoulos, *DWDM: Networks, Devices and Technology*, IEEE/Wiley, 2003.
3. S.V. Kartalopoulos, *Introduction to DWDM Technology: Data in a Rainbow*, IEEE/Wiley, 2000.
4. S.V. Kartalopoulos, *Next Generation Intelligent Optical Networks*: From Access to Backbone, Springer, 2008.
5. S.V. Kartalopoulos, "Next Generation Hierarchical CWDM/TDM-PON network with Scalable Bandwidth Deliverability to the Premises," *Optical Systems and Networks*, vol. 2, 2005, pp. 164–175, also online at http://www.sciencedirect.com
6. H. Hemmati, K. Wilson, M. Sue, D. Rascoe, F. Lansing, M. Wilhelm, L. Harcke, and C. Chen, "Comparative Study of Optical and RF Communication Systems for a Mars Mission, Part 1", Proc. SPIE, vol. 2669, Free-Space Laser Communication Technologies VIII, April, 1996.
7. H. Hemmati, J. Layland, J. Lesh, K. Wilson, M. Sue, D. Rascoe, F. Lansing, M. Wilhelm, L. Harcke, C. Chen, and Y. Feria, "Comparative Study of Optical and RF Communication Systems for a Mars Mission, Part 2", Proc. SPIE vol. 2990, Free-Space Laser Communication Technologies IX, May, 1997.

1

PROPAGATION OF LIGHT IN UNGUIDED MEDIA

1.1 INTRODUCTION

Free Space Optical communication (FSO) is found in a variety of telecommunication applications. In terrestrial applications, FSO technology employs a far infrared modulated beam through the atmosphere and particularly through its troposphere. In space applications, laser beams are used to establish inter-satellite links (ISL), so that a cluster of satellites forms a network in the sky [1]; in this case, satellites comprise nodes of a network, whereby the network may consists of geostationary (GS) or of low earth orbiting satellites (LEOS).

Either type of application, through troposphere or in space, owns a different set of issues [2]. In space the distance is very long and the satellite receivers may or may not face the sun, whereas in troposphere the medium is not homogeneous or stable. A third type of application comprises a stationary FSO node communicating with a moving FSO node, with additional tracking issues. As such, engineering FSO requires interdisciplinary expertise, so that the final FSO network provides high data rate links with reliability and at the expected performance under all or most environmental conditions in a profitable manner; the laser beam and the medium it travels through are two important entities of the FSO network.

Free Space Optical Networks for Ultra-Broad Band Services, First Edition. Stamatios V. Kartalopoulos.
© 2011 Institute of Electrical and Electronics Engineers. Published 2011 by John Wiley & Sons, Inc.

1.2 LASER BEAM CHARACTERISTICS

1.2.1 Wavelength

The laser beam that is used in FSO links has a wavelength of either 800 nm, 1310 nm, or 1550 nm. The mast popular of the three is 1550 nm for the following reasons:

- The 800 nm is generated by low cost vertical cavity surface emitting lasers (VCSEL) laser technology but the beam has low power and therefore the beam is modulated at very low data rates, up to 100 Mb/s and for link lengths of few hundred meters.
- The 1310 nm used to be a popular wavelength because of the distributed feedback (DFB) and Fabry-Perot type lasers, which support higher power than the VCSEL and therefore higher data rate and/or longer link lengths.
- The 1550 nm has been the most popular of all because it supports higher power levels, Gb/s data rate, longer link lengths, and also wavelength division multiplexing (WDM) technology [3, 4]; that is, several wavelengths in the 1520–1570 nm range and ITU-T standard compliant [5–8]; that is, an aggregate data rate which is the product of the number of different optical channels in the beam times the data rate in each channel. In some WDM applications, the 1310 nm is multiplexed with the 1550 nm to provide a two-channel WDM beam; this is acceptable in applications that do not require a large aggregate data rate; in addition, the 1310 and 1550 nm channels have large channel separation that turns out to be beneficial and convenient in receiver filter design. Moreover, a long-pass optical filter at the receiver rejects most wavelengths below the 1300 nm and it greatly reduces the solar background radiation (SBR) interference. The sun's photosphere emits electromagnetic radiation in a wide spectrum that is centered at a wavelength of 500 nm (the visible spectrum is from 400 to 700 nm) and at an average temperature in excess of 5500°C; the sun's radiation also includes wavelengths in the radio, Ultra Violet (UV), X-ray and Gamma-ray bands.

1.2.2 Beam Profile and Modes

As the laser beam emerges from the device, the intensity distribution along its cross-section is not uniform but it usually has a distribution; if the distribution is Gaussian, the beam is also termed "Gaussian".

The cross-section profile of the beam is of importance; in general, it is supposed to be circular with a uniform Gaussian distribution of 360°. Typically, lasers that emit beams with a pure Gaussian distribution are operating on the *fundamental transverse mode*, or "TEM_{00} mode", Figure 1.1.

In general, the analysis of beam profile is complex and the Hermite-Gaussian equations are used to describe the beam modes, which are designated as "TEM_{mn}", where m and n are polynomial indices in the x and y directions. With pure Gaussian distribu-

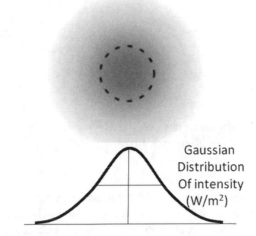

Figure 1.1. Intensity distribution of a Gaussian beam (cross section).

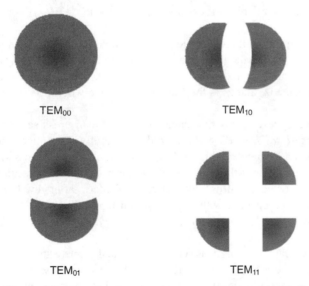

Figure 1.2. Hermite-Gaussian beam modes (approximated).

tion, m = n = 0 and thus TEM_{00}. Some lasers, however, are not as uniform and they operate in different modes, Figure 1.2.

1.2.3 Beam Divergence

Besides the non-uniformity of the beam cross section, the beam is not purely parallel in the z-direction (the direction of propagation). Even if the beam emerges parallel, it

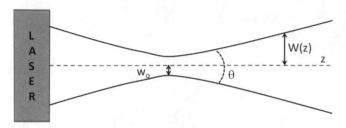

Figure 1.3. Divergent beam parameters: w_0 is the beam waist width w(z), depth of focus; Θ is the angular spread (angle of divergence after the waist).

does not remain so because of spatial diffraction, which causes the beam to first narrow at a point known as "waist", w_0, and then diverge at an angle Θ, Figure 1.3.

A laser beam propagating in the z direction and with Gaussian distribution across the beam is mathematically described by:

$$E(r) = E \cdot [w_o/w(z)] \cdot \exp[-r^2/w^2(z)] \cdot \exp[-jkz - jk(r^2/2R(z)) + j\tan^{-1}(z/z_o)] \quad 1.1$$

where E is the amplitude of the electric field, w_o is the minimum beam waist where the phase is constant, w(z) is the beam waist at distance z, r is the radius of the waist at z, k is the wave number approximated to $(k \sim 2\pi / \lambda)$, R(z) is the radius of the wave curvature at distance z, and z_o is the Rayleigh distance where the beam has expanded to $\sqrt{(2w_o)}$.

Beam divergence expands the diameter of the beam cross-section over distance, known as *geometrical spreading*, which reduces rapidly the optical power density of the beam, known as *geometrical spreading loss*. Starting with a cross section of one or less millimeter diameter at the aperture of the laser device, as a result of geometrical spreading the laser beam will be few meters in diameter after few kilometers. Beams with negligible divergence, or with an almost constant radius over the axis of propagation z are known as *collimated beams*; an optical device known as *collimator* helps to accomplish this.

The geometrical spreading loss (GSL) for typical laser beams with surface area of the transmit aperture SA_T, at distance R where the receiver is with surface area of the receive aperture SA_R, with constant divergence angle θ, and assuming constant power distribution across the beam, is estimated at:

$$GSL = \frac{\text{Surface area of receive aperture}}{\text{Surface area of beam at distance R}} = \frac{SA_R}{SA_T + (\pi/4)(\theta R)^2} \quad 1.2$$

Typical values for SA_T and SA_R are in the range 0.004–$0.025 \, m^2$, for an angle θ of 1 to 3 mrad.

The aforementioned spreading loss is for an ideal beam and also a good approximation. However, practical beams do not have a constant power distribution across the beam but they may have a Gaussian distribution; as aforementioned, laser beams are

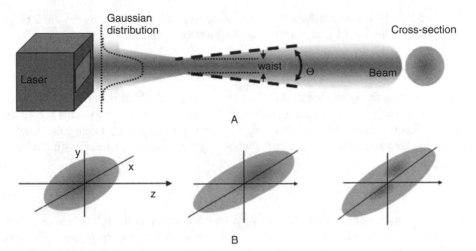

Figure 1.4. Gaussian distribution of circular cross-section and non-Gaussian irregular power distribution.

not purely conic but they may have a longitudinal profile with a waist of radius W_0, after which they diverge at an almost constant angle. In calculations, one may use the full width at half-maximum (FWHM), which is provided in the manufacturer's data sheets; the FWHM, or 50% of intensity, is $0.59W_0$.

Additionally, in practical beams the cross-section and the intensity distribution may be neither circular nor Gaussian, Figure 1.4; non-circular cross-section reduces the coupling efficiency of the beam onto the fiber and onto the FSO receiver due to irregular power distribution and irregular divergence. For non-Gaussian profiles, an integral formula may be used for waist and radius calculations.

The optimum waist, $w_{0,optimum}$, for a wavelength λ at a distance z from the source is defined as:

$$w_{0,optimum} = [\lambda z/\pi]^{1/2} \qquad\qquad 1.3$$

Similarly, the distance at which the radius spreads by a factor $\sqrt{2}$ is called Rayleigh range, z_R, and it is defined by:

$$z_R = \pi w_0^2 / \lambda \qquad\qquad 1.4$$

The *beam parameter product* (BPP) is a measure of laser beam quality or its proximity to an ideal Gaussian beam. BPP is the product of waist radius W_0 and the far field divergence of the beam. The deviation of a beam from a Gaussian beam at the same wavelength is indicated by a factor, M^2; this factor is the ratio of the actual beam BPP to an ideal beam BPP at the same wavelength.

For example, pure Gaussian beams have a value of $M^2 = 1$, He-Ne laser beams are very close to Gaussian with $M^2 < 1.1$, and diode laser beams have M^2, from 1.1 to 1.7.

Finally, when divergent beams are corrected to parallel with a lens system, known as *collimator*, then the beam is known as *collimated beam*.

1.2.4 Rayleigh Range

The point from the beam waist where the beam area is doubled is known as *Rayleigh length* or *Rayleigh range*; actually, this is the distance from the beam waist where the beam radius is increased by $\sqrt{2}$. For Gaussian beams, the Rayleigh length, Z_R, is determined by the waist radius W_0 and by the wavelength λ (assuming a beam with moderate divergence):

$$Z_R = (\pi W_o^2)/\lambda \qquad\qquad 1.5$$

where the wavelength λ is the vacuum wavelength divided by the refractive index n of the medium in which the beam propagates. The Rayleigh length is decreased by a factor M^2 for beams with non-Gaussian profile. In either case, notice that Z_R depends on the wavelength λ. Twice the Rayleigh length is known as the *confocal parameter b* [9–11].

In free space propagation of Gaussian beams, the beam cross-section or spot size at a distance z, $W(z)$, is expressed in terms of the waist and the Rayleigh length:

$$W(z) = W_o \sqrt{\{1 + (z/Z_R)^2\}} \qquad\qquad 1.6$$

where the origin on the z axis is approximated at the waist point.

Similarly, the width of the beam at the Rayleigh distance Z_R is $W_o\sqrt{2}$, and the confocal parameter, b, is:

$$b = 2 Z_R = (2\pi W_o^2)/\lambda. \qquad\qquad 1.7$$

For points far away from the Rayleigh range, $z \gg Z_R$, the divergence is approximated to:

$$\theta \sim \lambda/(\pi W_o), \qquad\qquad 1.8$$

and the total angular spread of the beam to:

$$\Theta = 2\theta. \qquad\qquad 1.9$$

Because of these approximations, the Gaussian beam model holds for waists larger than $2\lambda/\pi$.

Geometrically, beam divergence or spreading is the derivative of the beam radius with respect to the distance from the beam waist. Beam divergence has an important effect at the receiver: the optical power per square unit area of the beam cross-section weakens fast as it propagates, and because the receiving photodetector has a small detecting area, a fraction of a square centimeter or few square millimeters, the optical power that impinges onto the photodetector is very small and the remaining power of the laser beam spills over, Figure 1.5.

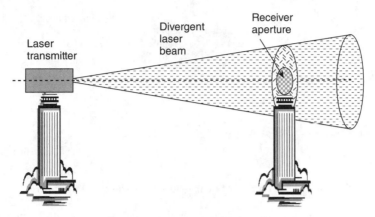

Figure 1.5. Laser beam spill over.

Thus, from a receiver viewpoint, geometrical spreading is equivalent to power attenuation because the beam spreads continuously as it travels; thus, the *geometrical spreading attenuation* (GSA) depends on beam divergence (angle) and link length, and it is defined as:

GSA = Aperture of receiver/Beam cross-section area at the receiver

In general, small beam divergence is preferred because beams with large divergence require complex optical lens without spherical aberrations.

1.2.5 Near-Field and Far-Field Distribution

The beam generated by a laser device is not perfectly narrow, or cylindrical, or centered, and the almost Gaussian distribution in the x-axis may differ from the distribution in the y-axis. In addition, the beam intensity distribution at the "edge" or the output facet of the laser device (known as the *aperture of the source*) is not the same with the intensity distribution at some distance. At a short distance from the source aperture the intensity distribution is known as *near-field*, and at a far distance where the intensity distribution seems to remain almost unchanged is known as *far-field*, Figure 1.6.

The near-field region is where light rays exhibit disorder phase fronts. This region is also known as the *Fresnel zone*. The far-field region is where fronts have become ordered and the beam propagation characteristics have been stabilized. This region is also known as *Fraunhofer zone*. Because the terms "near" and "far" are subjective, a metric has been developed to distinguish between the two. Thus, the near-field distance, D_{nf}, and far-field distance, D_{ff}, have been expressed in terms of the source aperture (the area of the laser waveguide at the edge) and to the wavelength of the laser light, as:

$$D_{ff} \gg \pi d^2 / \lambda \qquad\qquad 1.10$$

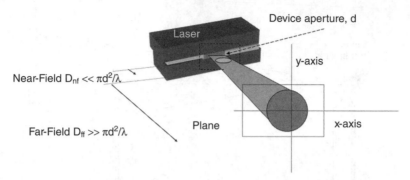

Figure 1.6. Definition of near field and far-field of laser beams.

and

$$D_{nf} \ll \pi d^2/\lambda \qquad\qquad 1.11$$

Which of the two parameters (D_{ff} or D_{nf}) is more suitable depends on optical design requirements. For monochromatic light, and if focusing lens are incorporated in the design, then the far-field is better suited. In addition, the far-field angular dispersion is superior to the near-field. If on the other hand, the laser light needs to be coupled in another waveguide in the vicinity of the laser, the near-field may be better suited. For laser diodes with divergent beams, the near-field is few microns from the output facet. If the far-field divergence angle is θ, and the near-field width is w, a beam parameter for a given wavelength λ, K, is defined as:

$$K = 4\lambda/\pi w\theta \qquad\qquad 1.12$$

Occasionally, a similar parameter M is used, defined as $\sqrt{M} = 1/K$. Typically, laser manufacturers provide near- and far-field data for their devices.

1.2.6 Peak Wavelength

Laser devices do not source a purely monochromatic beam. In practice, they source a continuum of wavelengths in a narrow spectral range with a near-Gaussian distribution. The wavelength with the highest radiant intensity of the source is known as *peak wavelength*. The wavelength spread at either side of the peak is measured in nm; e.g., 1550 +/− 2 nm indicates that 1550 is the peak wavelength with a Gaussian spread at either side of 2 nm.

A *monochromator* is a narrow band-pass filtering device that allows a very narrow spectral band to pass through it.

1.2.7 Degree of Coherence

Laser beams, by definition may be considered coherent; that is, the wavefront of all rays departing the laser device are in phase. However, this is not absolutely true. For example, in simple interference experiments, when the intensity minima and maxima of the fringes in the interference pattern are well defined and crisp, the beam is *coherent*, and when they are not (the pattern appears blurry), the beam is *incoherent*.

The *degree of coherence* (DoC) corresponds to the percent of rays in phase in the beam. For example, if in an interference experiment the minimum and maximum intensity of the fringed pattern is I_{min} and I_{max}, respectively, then the degree of coherence is defined as:

$$DoC = (I_{max} - I_{min})/(I_{max} + I_{min}) \qquad\qquad 1.13$$

In general, a beam is considered coherent when the degree of coherence is above 0.88, partially coherent if it is less than 0.88 but above 0.55, and incoherent if it is <0.5.

Notice that, although a sourced beam may start being coherent, as the beam travels through matter with non-uniform dielectric, DoC may change. Thus, the length of travel during which the beam remains coherent (>0.88) is known as the *coherence length*, and correspondingly, the travel time along this length is known as the *coherence time*.

When coherency of two interacting rays varies with time, then a time varying interferogram is produced known as a *speckle pattern*. In FSO systems, this in fact gives rise to the phenomenon of *scintillation* [12].

1.2.8 Photometric Terms

For comparison between two light sources or two illuminated objects, the following measurable units are introduced.

- (Total) *luminous flux*, Φ, is the rate of optical energy flow (or number of photons per second) emitted by a point light source in all directions; it is measured in lumens (lm). In radiometric terms this is known as optical power and it is measured in Watts.
- *Luminous* (or *candle*) *intensity*, I, is the rate emitted in a solid angle of a spherical surface area equal to its radius (e.g. radius = 1 m, surface area = 1 m^2). Luminous intensity is measured in *candelas* or *candles* (cd). The luminous intensity of a sphere is $\Phi/4\pi$.
- *Illuminance*, E, is the flux density at an area A (m^2), or the luminous flux per unit area; it is measured in *lux* (lx). The illuminance at a point of a spherical surface is $E = \Phi/4\pi R^2$. Illuminance refers to light received by a surface. Since the luminous intensity I of the sphere is $\Phi/4\pi$, then $E = I/R^2$. This is known as the *law of inverse squares*, Figure 1.7.
  - *Luminance*, B, is the amount of optical energy emitted by a lighted surface per unit of time and per unit of solid angle and per unit of projected area is known as. Luminance is measured in cd/m^2 and is also known as nit (nt).

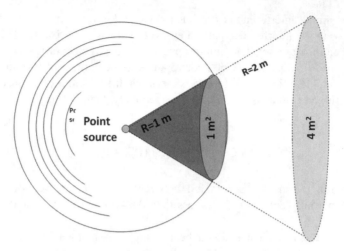

Figure 1.7. Illustration of the law of inverse squares.

Some examples of luminance are (in cd/m^2):

Clear blue sky: 10^4
Sun: 1.6×10^9
Candle: 2×10^6
Fluorescent lamp: 10^4

Table 1.1 summarizes the photometric units that are used in optics and optical communications, their measuring units and their dimensions (M = mass. T = time, L = length):

1.2.9 Radiometric Terms

Similarly, there are radiometric terms defined as follows:

- *Radiant power* (φ) or optical power is the rate of flow of radiant energy and it is measured in Watts (W).
- *Radiant energy* (Q) is the energy transferred by electromagnetic waves. It is the time integral of radiant power and it is measured in joules (J).
- *Radiant intensity* (I) is the radiant power per unit solid angle and it is measured in Watts/steradiant (W/sr).
- *Radiance* (L) in a given direction is the radiant intensity per unit of projected area of the source as viewed from that direction, and it is measured in W/sr-m^2.

TABLE 1.1. Photometric Units (M = mass. T = time, L = length)

Definition	Photometric unit	Dimensions
Energy:	Luminous energy (talbot)	ML^2T^{-2}
Energyper unit area:	Luminous density (talbot/m^2)	MT^{-2}
Energyper unit time:	Luminous flux (lumen)	ML^2T^{-3}
Flux per unit area:	Luminous emittance (lumen/m^2 or lambert)	MT^{-3}
Flux per unit solid angle:	Luminous intensity (lumen/steradian)	ML^2T^{-3}
Flux per unit solid angle per unit projected area:	Luminance (candela/m^2)	MT^{-3}
Flux input per unit area:	Illuminance (meter-candela)	MT^{-3}
Ratio of reflected to incident flux:	Luminous reflectance	
incident flux:	Luminous transmittance	
Ratio of absorbed to incident flux:	Luminous absorptance	

- *Spectral radiance* (L_λ) is the radiance per unit wavelength interval at a given wavelength, and it is measured in Watts/steradiant-unit area-wavelength interval (W/sr.m^2.nm).
- *Irradiance* (E) or power density is the radiant power per unit area incident upon a surface, and it is measured in Watts/square meter (W/m^2).
- *Spectral irradiance* (E_λ) is the irradiance per unit wavelength interval at a given wavelength, and it is measured in Watts/unit area-unit wavelength interval (W/m^2.nm).

1.2.10 Beam Power and Intensity

The optical power P(r, z) through a circular hole of radius r in the transverse plane at point z of the propagation axis is calculated from:

$$P(r,z) = P_o[1 - \exp\{-2r^2/(w^2(z))\}] \qquad 1.14$$

where P_o is the power of the beam emitted by the laser:

$$P_o = \pi I_o W_o^2/2. \qquad 1.15$$

where I_o is the peak intensity of the beam.

The above relationship is approximated for a circle of radius r = w(z) as

$$P(r,z) = (1 - e^2) P_o = 0.865 P_o. \qquad 1.16$$

Thus, if the radius is r = 1.224 w(z), then 95% of the beam power will flow through the circle.

Similarly, the peak intensity I(z) at a distance z from the beam waist is twice the average intensity, which is obtained by dividing the total power emitted by the laser by the area within the radius w(z).

Example: Consider a laser beam emanating from a laser device with 260 mW power, a spot size (beam cross-section) 1.5 mm diameter, and with Gaussian distribution profile. For a beam with a divergence angle 1.5 mrad, the light intensity or irradiance (optical power/area of beam cross-section) at 3 meters is ~9.2 mW/mm^2. If divergence is reduced by 20% (down to 1.2 mrad), the irradiance at 3 meters is ~12.7 mW/mm^2, that is an improvement of ~40%. This means that, as the angle of beam divergence decreases, the light intensity increases exponentially and thus the same laser beam is useable over longer distance.

1.2.11 The Decibel Unit

Optical power, power attenuation and optical loss are measured in decibel units (dB). Power in decibel units is defined as ten times the logarithm (base 10) of power (in Watts):

$$\text{Power (dB)} = 10\log_{10}[\text{P (Watts)}] \qquad\qquad 1.17$$

In communications, the transmitted optical signal is at extremely low power, in the order of milliwatts. To denote this, the decibel unit is expressed in dBm:

$$\text{Power (dBm)} = 10\log[\text{P (mWatts)}] = 10\log[\text{PX10}^{-3}\text{ (W)}]$$

Here, few properties of logarithms are reminded, Table 1.2, as they play a key role in the understanding of dos and don'ts when dealing with decibels.

TABLE 1.2. Properties of Logarithms

---

1. $\log(AB) = \log A + \log B$
2. $\log(A/B) = \log A - \log B$
3. $\log(A^N) = N \log A$
4. $\log \{N\text{root}(A)\} = (1/N)\log A$
5. $\log 10 = 1$
6. $\log 1 = 0$
7. $\log A = \log(e) \ln A$
8. $\log (e) = \log(2.71828+) = 0.434294$
9. $\ln 10 = 2.30258+$
10. $\ln 2.71828 = 1$
11. $\log A = +\log A$, where $A > 1$
12. $\log A = -\log|A|$, where $0 < A < 1$
13. $\log(A + B)$ *not equal to* $\log A + \log B$

When adding/subtracting dBs and dBms, caution should be taken regarding the mix-and-match of units. Decibel units are additive if their argument is multiplicative, and therefore, if the units are not handled correctly one may make a serious erroneous calculation.

Power attenuation or power loss is also expressed in decibel units. In this case, it is (ten times) the logarithm of the ratio of received power over transmitted power, both expressed in the same units and thus the ratio is dimensionless. Hence, the attenuation over a fiber span is expressed in dB, regardless whether the power is in W or mW units):

$$\alpha(\lambda) = 10 \log P_1/P_2 \ (dB) \qquad\qquad 1.18$$

As an example, a power ratio of 1000 is 30 dB, of 10 is 10 dB, of ~3 is 5 dB, of 2 is ~3 dB, and a ratio of 0.1 is −10 dB. To the untrained eye and mind, loss or gain in dB's is very abstract as we are trained to think of loss and gain in ratios, such as for example, 100 times less or 100 times more versus 20 dB. However, the dB is a more convenient unit and in optical communications it is widely used.

Besides decibels, power loss is also provided as a percent of power transmitted compared with that received, such as 60%, and so on. That is, if 100 units of power were transmitted and 60 were received, the loss is 100–60 = 40 and in percent it is (100–60)/100. The correspondence of dB to percent is easy to calculate. For example, 90% power loss corresponds to $10\log\{(100-90)/100\} = -10$ dB, 50% corresponds to $10\log 0.5 = -3$ dB and 2% to $10\log 0.98 = -0.01$ dB. Table 1.3 lists conversions from dB loss to % loss and from % loss to dB loss.

A power ratio that is widely used in communications is the signal-to-noise (SNR), also expressed in dB units.

1.2.12 Laser Safety

The optical spectrum used in communications is invisible to the human eye. Although the eye has high absorbance in 1550 nm, nevertheless the general rule is that it is very

TABLE 1.3. Conversion from dB loss to % and from % to dB loss

dB loss	%loss	%loss	dB loss
0	0	0	0
−0.1	−2.3	−0.5	−0.02
−0.5	−10.9	−1	−0.04
−1	−20.6	−5	−0.22
−2	−36.9	−10	−0.46
−5	−68.4	−40	−2.22
−10	−90.0	−90	−10.00
−20	−99.0	−99	−20.00

risky to look straight into a laser beam or into a lit fiber; this invisible light may be at a power level or at an irradiance that can permanently damage the *cornea* and/or the *retinal sensors*. Retinal sensors do not regenerate and once they are damaged they remain so.

The eye physiology is such that an image is focused onto the *retina* where there are about 130 millions of light sensors, rods and cones, the axons of which send electrochemical signals to other retinal neurons where image preprocessing takes place, and from where other signals are sent via the optic nerve to the brain for final post-processing.

Now, because the eye automatically focuses light, a beam with cross section $1\,cm^2$ on the lens is concentrated to less than $20\,\mu m^2$ onto the fovea. This represents an enormous power density factor, such that $1\,W/cm^2$ on the lens becomes many kW/cm^2 on the fovea, which may damage it permanently.

There are two factors to consider in retinal damage, radiance of the beam and time of exposure (continuous or pulsed beam), and one safety rule, use laser eye protectors. A source is considered continuous if it emits light continuously from 0.25 to 30,000 seconds; below 0.25 sec, it is considered pulsed, and ordinary eye protectors made with polycarbonate can withstand irradiances up to $100\,W/cm^2$ ($1\,MW/m^2$) at $10.6\,\mu m$ for several seconds.

Federal law mandates affixing the classification of the laser source on the device.

1.2.13 Classification of Lasers

The transmitter and specifically the light source is one of the key components in optical communications systems. Light sources must be compact, monochromatic, stable, and have a long lifetime (many years). Stability implies constant optical power level and wavelength (over time, voltage and temperature variations), that is, no power variation and wavelength drift. In practice, because there are no absolute monochromatic light sources, a very narrow band of wavelengths with a Gaussian distribution is desirable.

Light sources are classified as *coherent* (when all emitted photons are in phase) and *incoherent* (when emitted photons have random phase).

The first classification includes all lasers and the second includes light emitting diodes (LED) and incandescent sources.

Light sources are also classified as *continuous wave* (CW), and as directly *modulated*.

In communications, CW sources require modulators that are placed in the optical path. In this arrangement, an electrical signal representing a data stream acts upon the modulator affecting the continuous flow of light. Modulators are affected by the application of a modulated voltage, current, or light. They may be external or integrated with the laser device, Figure 1.8.

Based on the maximum *accessible emission limits* (AEL), or the driving optical power (in Watts), or the energy (in Joules) by wavelength and exposure time, light sources had been classified in four classes, from class 1 (no hazard when used in

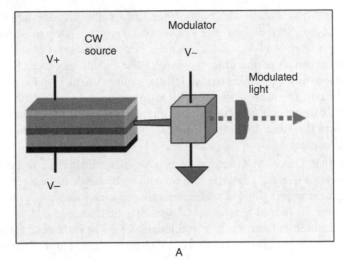

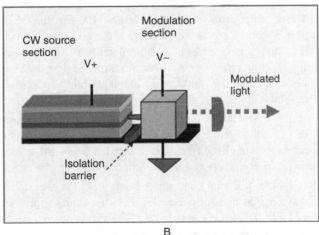

Figure 1.8. A light modulator can be external (A) or integrated with the laser device (B); CW = continuous source.

normal ways) to class 4 (hazardous to eye and skin). Because since 1970 laser technology has been advanced and laser applicability has been expanded, the classification system has been revised (IEC 60825–1, edition 2.0, March 3, 2007, is applicable to safety of laser products emitting laser radiation in the wavelength range 180 nm to 1 mm).

Under the old system, in the United States the class numbers were the Roman numerals (I to IV), and in the European Union the Arabic numerals (1–4). The revised system uses numerals (1–4) in all jurisdictions. The revised four classes and their subclasses are:

- **Class 1:** This class includes high-power lasers within an enclosure that prevents exposure to the radiation; the enclosure cannot be opened without shutting down the laser. A class 1 laser is safe under all conditions in normal use; that is, the maximum permissible exposure (MPE) is not exceeded. For example, a continuous laser at 600 nm (visible) can emit up to 0.39 mW, but for shorter wavelengths the maximum emission is lower because potentially those wavelengths can generate photochemical damage. The maximum emission is also related to the pulse duration, in the case of pulsed lasers, and to the degree of spatial coherence.

- **Class 1M:** Class 1M lasers produce large-diameter beams, or beams that are divergent. A Class 1M laser is safe for all conditions of use except when passed through magnifying optics such as microscopes and telescopes. If the beam is refocused, the hazard of Class 1M lasers may be increased and the product may be reclassified. A laser can be classified as Class 1M if the total output power is below class 3B but the power that can pass through the pupil of the eye is within Class 1.

- **Class 2:** Class 2 applies to visible-light lasers (400–700 nm). Class-2 lasers are limited to 1 mW continuous wave, or more if the emission time is less than 0.25 seconds, or if the generated light is not spatially coherent. Because the *blink reflex* of the eye limits the laser exposure to no more than 0.25 seconds, Class 2 laser is safe, unless the the blink reflex is intentionally suppressed. Many *laser pointers* are class 2.

- **Class 2M:** As with class 1M, this class applies to laser beams with a large diameter or large divergence, for which the amount of light passing through the pupil cannot exceed the limits for class 2. Thus, a Class 2M laser is safe because of the blink reflex if not viewed through optical instruments.

- **Class 3R:** Visible spectrum continuous lasers in Class 3R are limited to 5 mW; for other wavelengths and for pulsed lasers, other limits apply. A Class 3R laser is considered safe if handled carefully, with restricted beam viewing. With a class 3R laser, the MPE can be exceeded, but with a low risk of injury.

- **Class 3B:** A Class 3B laser is hazardous if the eye is exposed directly, but diffuse reflections such as from paper or other *matte* surfaces are not harmful. Continuous lasers in the wavelength range from 315 nm to far infrared are limited to 0.5 W. For pulsed lasers between 400 and 700 nm, the limit is 30 mJ. Other limits apply to other wavelengths and to *ultrashort pulsed* lasers. Protective eyewear is typically required when directly viewing a Class 3B laser beam. Class 3B lasers must be equipped with a key switch and a safety interlock.

- **Class 4:** Class 4 lasers include all lasers with beam power greater than Class 3B. By definition, a Class 4 laser can burn the skin, in addition to potentially causing permanent eye damage as a result of direct or diffuse beam viewing. These lasers may ignite combustible materials, and thus may represent a fire risk. Class 4 lasers must be equipped with a key switch and a safety interlock. Most entertainment, industrial, scientific, military, and medical lasers are in this category.

Because many lasers in use were manufactured before the revised classification, the old classification is [found in 13]:

- **Class I:** Class I includes all lasers for which the power is low so that there is no eye damage even after hours of exposure. This includes more hazardous laser devices, which however are in enclosures preventing user access to the laser beam during operation, such as CD players. Laser sources must satisfy Class I Laser Safety requirements according to the US Food and Drug Administration (FDA/CDRH) and international IEC-825 standards.

- **Class II:** Class II are low power laser sources that dot impose hazard due to the blinking of the eye. However, prolonged exposure may cause damage. Class II are lasers emitting light in the visible spectrum (0.4 to 0.7 mm) at an average power up to 1 mW power, and if applicable a pulse duration of less than 0.25 seconds. The blink reflex of the human eye (known as *aversion response*) will prevent eye damage An example of a Class II laser is a HeNe laser pointer of 1 mW or less.

- **Class IIa:** Class IIa are lasers emitting light in the visible wavelengths (0.4 to 0.7 mm) which produce a burn to the retina if directly and continuously viewed in excess of 1000 seconds. Supermarket laser scanners are in this subclass.

- **Class IIIa:** Class IIIa laser sources when viewed for <0.25 seconds are not hazardous. However, they may be if the exposure is prolonged or if the laser beam is focused by a lens. Class IIIa are medium power lasers 1 to 5 mW and they represent a potential hazard to the eye. Beam power density may not exceed 2.5 mW/square cm. Class IIIa includes lasers with an accessible output power between 1 to 5 times the Class I AEL for wavelengths shorter than 0.4 μm or longer than 0.7 μm, or less than 5 times the AEL for wavelength between 0.4 and 0.7 μm. Lasers applicable to *firearms* and *laser pointers* are in this category.

- **Class IIIb:** Class IIIb laser sources may cause damage if viewed directly or through specular reflections but do not produce hazardous reflections. Class IIIb lasers are typically less than 0.5 W average power. The hazard for IIIb lasers is potentially greater than that for IIIa. The hazard is still limited to direct viewing of the laser beam. These lasers do not produce hazardous diffuse reflections or represent a skin exposure hazard. Protective eyewear is recommended when direct beam viewing. Lasers at the high power end of Class IIIb may also present a fire hazard and can lightly burn skin.

- **Class IV:** Class IV laser sources cause damage if viewed directly or through specular reflections and produce hazardous reflections. Class IV lasers are at 0.5 W and higher. These lasers represent hazards (eye damage, skin injury, and or potential flammable material ignition source) for direct viewing, viewing of diffuse reflections, and skin exposure. Most entertainment, industrial, scientific, military, and medical lasers are in this category.

Note: Multiwave lasers are classified (with the old or the revised system) according to the most hazardous wavelength or the most hazardous possible wavelength configuration.

In general, laser sources in communications are of class 1 or class I with Hazard Level 3A. System designers and fiber connector designers take precautions so that there is automatic power laser shutdown (APSD) as soon as a fiber is disconnected from the source and particularly where laser power radiation exposure is greater than 50 mW (17 dBm). Standards that describe recommendations for laser safety are:

- The ITU-T recommendation G.664 provides optical safety procedures.
- The American National Standard Institute (ANSI Z136.1-2000) provides maximum permissible exposure (MPE) limits and exposure durations, from 100 fempto-seconds to 8 hours.
- Similarly, the American Conference of Governmental Industrial Hygienists (ACGIH) provides the threshold limit values (TLV) and biological exposure indices (BEI).
- Both ANSI and ACGIH have become the basis for the U.S. Federal Product Performance Standard (21 CFR 1040).

In summary, depending on intensity of the source, laser sources in the range 700–1400 nm may cause retinal burn and promote cataract, whereas lasers in the range 1400–3000 nm may cause corneal burn, may affect the protein of the aqueous humour of the eye, and may promote cataract; the optical communications spectrum is in the range 1280–1620 nm, whereas the most used wavelengths in FSO are in the range 700–800 nm and in the range around 1550 nm.

1.3   ATMOSPHERIC LAYERS

The atmosphere is retained by the Earth's gravitational field and it consists of a mixture of gases that form a layer surrounding Earth [14]. The overall atmosphere consists approximately of (by volume) 78.00% Nitrogen, 21% Oxygen, 1% Argon, 0.04% Carbon dioxide, and other gases in smaller amounts (He, Ne, CH_4, H_2, Kr, and others), ~1% vapor, natural and manmade aerosols and pollutants (fluorine, chlorine, mercury, sulfur dioxide (SO_2), as well as dust, spores, pollen, and other; notice that this composition fluctuates within a year; the mass of atmosphere (or the total mean mass) is 5.1480×10^{18} kilograms (kg) with an annual range due to water vapor of about 1.5×10^{15} kg.

Because of the gravitational field, the vertical distribution of gases is not uniform, the density of gases is not equally distributed vertically from the surface of the earth, and the pressure is not the same at all heights. In addition, the temperature of the atmosphere varies with altitude. As a result, three quarters of the atmosphere's mass is within 11 Km from the surface. Because at different heights from the surface of the Earth there are different effects, the atmosphere is distinguished in layers (from lowest to highest), Figure 1.9:

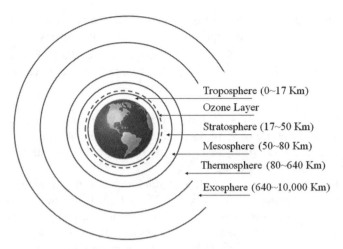

Figure 1.9: Layers in Earth's atmosphere.

1. **Troposphere:** it extends from the Earth's surface (sea level) and up to 7–17 Km; however, the thickness of this layer at the poles is about half of the thickness around the equator. The troposphere is the layer with the highest molecular density (approx. 80% of the Atmosphere's total mass), and also pressure (at sea level it is 1 atm = 760 torr = 101.3 KPa = 14.7 psi = 29.9 inches of Mercury); the air pressure decreases almost exponentially, approximately by 0.5 every 5.6 Km, or by $1 - 1/e = 1 - 0.368 = 0.632$ or 63% every 7.64 Km. The density of air at sea level is about 1.2 Kg/m^3. Typically, the *atmospheric density* decreases with altitude, z, and it is described by the *barometric formula*; this is an exponential formula that calculates the atmospheric pressure $P = P_o\exp(Mgz/RT)$ based on Earth's air mass, M, gravity, g, and temperature, T; R is a universal constant for air, 8.31432 N·m/(mol·K). The atmospheric density does not remain constant but it fluctuates during the day and thought-out the year. Most weather phenomena (rain, snow, fog, lightning, clouds, etc.) that affect daily life as well as FSO communications occur within the Troposphere.

2. **Stratosphere:** it extends from 7–17 Km to about 50 Km. The temperature increases with height. It also contains Ozone (O_3) at a concentration of few parts per million and in a layer that extends from 15 to 35 Km; this is known as the ozonosphere and it is created by the sun's ultraviolet (UV) light.

3. **Mesosphere:** it extends from 50 Km to 80–85 Km. The temperature decreases with height reaching $-100°C$. This is where most meteors burn up when entering the atmosphere.

4. **Thermosphere:** it extends from 80–85 Km up to 640 Km. It contains very low pressure particles. The temperature increases with height and it may reach 1500°C. The International Space Station orbits in this layer and between 320 and 380 Km.

5. **Ionosphere:** it extends from 50–1000 Km. It contains charged (ionized) particles and therefore it is important to radio communications.

6. **Exosphere:** it extends from 5000–1000 Km up to 10,000 Km. It borders the outer space and it contains few particles, which however cause atmospheric drag on satellites.

The interface between two successive layers have also names, such as that between the troposphere and the stratosphere is called *tropopause*, between the stratosphere and the mesosphere is called *stratopause*, and the one between the mesosphere and the thermosphere is called *mesopause*.

Notice that the aforementioned layers (thickness and positioning) vary continuously within a day and their borders overlap. The actual air pressure, density, temperature and molecular consistency in each depend on the relative position of moon and sun with respect to the Earth's surface, on solar activity, and on many other factors.

In the following section, we focus on the Troposphere and particularly on the phenomena that affect FSO communications.

1.4 ATMOSPHERIC EFFECTS ON OPTICAL SIGNALS

FSO technology in terrestrial applications depends on the propagation of the laser beam through the troposphere. The troposphere is the layer in which many weather phenomena occurs, which interact with and affect the quality of the propagating optical signal.

Although weather predictions can be made with the help of radars and weather satellites, overall, the troposphere is a highly dynamic and unstable medium; it consists of gases, vapor, airborne dust, natural and manmade aerosols, pollutants, and other particles, which with sunlight, temperature and pressure variation continuously move and change their consistency and characteristics so that a dynamic model of the troposphere is highly complex and difficult to construct [15, 16].

All molecules and particles in the troposphere interact with light and, in addition to chemically interacting, they cause absorption, scattering, fog attenuation, rain attenuation, snow attenuation, scintillation, lightning discharges, atmospheric tides, and other effects that affect the FSO optical propagating signal.

As such, in free space optical communication, the troposphere as propagation medium should be examined and understood. The understanding of tropospheric phenomena and how they affect the propagating light helps to better engineer effective, intelligent and cost-efficient FSO links and reliable networks with the ability to self-adjust laser emission power and receiver sensitivity, self-control link alignment, self-balance network traffic load, and self-avoid affected areas in order to provide uninterrupted service at the expected quality.

In this section, we examine the most important tropospheric phenomena that affect FSO communications. However, because the refractive index of air is an important parameter that affects light propagation through it, we will start with it.

1.4.1   Refractive Index of Air

The *index of refraction* of the atmosphere has been an important parameter in the study of light propagation through it; the study of the index of refraction of air is highly complex due to the many variables that continuously change under dynamic conditions, some predictably and some unpredictably. Thus, some studies started as early as 1700, predominantly motivated by astronomical observations. For example, differential refraction due to atmospheric refractive index variation, and depending on the wavelength of observation, causes the image of an object to appear at different positions in the focal plane of a telescope.

With the advent of technology and particularly the computer, last century witnessed a significant impetus in modeling the index of refraction of the air. Although the purpose of this section is not to provide a complete historical overview of the developed models on the refraction index of air, it is of interest to briefly look into some models, as they are still in use and because FSO is a beneficiary of a large body of research in atmospheric refractive index modeling [17].

In 1939, Barrell and Sear [18] formulated the first mathematical model of the refractive index of air in the visible spectrum at T = 0°C and P = 452 mmHg.

In 1953, Edlén formulated an empirical model by fitting a curve on data for the refractive index of "standard air" at a vacuum wavelength of λ_{vac} (μm); that is, air at temperature T = 15°C (288.15°K), atmospheric pressure P = 760 mmHg (1013.25 mbar), and containing 300 ppm CO_2 [19], which is:

$$(n-1)\times 10^8 = 6432.8 + 2949810/(146 - \lambda^{-2}) + 25540/(41 - \lambda^{-2}) \qquad 1.19$$

where n is the refractive index, and λ is the wavelength (in μm) in vacuum.

In 1966, this model was revised and it was corrected; after re-fitting the data, the number 41 in the denominator was replaced by the number 38.9 [20].

In 1967, Edlén's formula was revised by Owens [21] who consired air compressibility effects deviating from the ideal gas.

In 1972, Peck and Reeder [22] made new measurements in the infrared region (IR), with a refractive index accuracy of 2 parts in 10^9 in the region 230 nm (UV) to 1.69 mm (near IR); FSO is within this region because it uses wavelengths at ~800 nm and 1.55 mm.

In 1981, Jones reconsidered the real-gas and formulated his own model [23]; subsequent to this, more models have been formulated [24–31] in an attempt to consider water and CO2 in the atmosphere, and also other parameters that affect the refractive index of air in an attempt to provide the most accurate model.

1.4.2   Atmospheric Electricity

Atmospheric electricity, between clouds and between cloud and earth's surface, builds up a typical voltage differential that ranges from 20,000 Volts to 100 Mega-Volts, which can cause multiple lightning discharges through the atmosphere, each up to 35,000 Amperes.

Lightning emits flashes of electro-magnetic waves from very long to very short wavelength, such as radio waves (RF, VHF, UHF), optical, x-rays, and up to gamma rays. The plasma temperature in the lightning can reach 28,000°K and the electron density may exceed $10^{24} e^-/m^3$ [32]. The duration of each flash is between 20 to 130 msec; in FSO communications at 1 Gb/s, 100 msecs flash is equivalent to 10^8 bits (or 12.5 Mbytes), which corresponds to substantial data loss if the flash interferes with the signal at the receiver.

1.4.3 Atmospheric Tide

Vapor and ozone absorb the Sun's periodic radiation causing atmospheric tides in the Troposphere and in the Stratosphere. Atmospheric tides propagate in an atmosphere where density varies significantly with altitude causing wind, temperature, density and pressure fluctuations with a periodic oscillation of about 24 hours. Near ground level, atmospheric tides are manifested with semidiurnal pressure minima (at 4 am and 4 pm local time) and pressure maxima (at 10am and 10 pm local time) [33].

1.4.4 Definitions

1.4.4.1 Parts Per Million by Volume (PPMV)   PPMV of an element in the air is the number that expresses the fraction of total molecules of a gas element in the unit of volume of air or by mole of air and it refers to the concentration of a gas element in the air [34]. For example, 1 PPMV is a micro-liter (10^{-6} liter) of a specific gas that is mixed in a litter of air.

Another unit for the concentration of a gas element in the air is the metric unit milligram per cube-meter ($\mu g/m^3$). To convert PPMV to $\mu g/m^3$ one needs the density of the particular gas; this is calculated using Avogadro's Law that stipulates: *equal volumes of gases, at the same temperature and pressure, contain the same number of molecules*. Avogadro's Law implies that 1 mole of gas at *standard temperature and pressure* (STP) has a volume of 22.71108 liters; this is also known as the molar volume of ideal gas [35]. The Internet provides PPMV to metric units converter tools, one of them found at http://www.lenntech.com/calculators/ppm/converter-parts-per-million.htm.

1.4.4.2 Visibility   This is a term specifically defined for meteorology, aviation and traffic in general. It characterizes the degree of transparency of the atmosphere in the visible spectrum; that is, as is seen by a human observer.

Visibility is measured by the *runway visual range* (RVR); that is, the distance a luminous parallel beam travels through the atmosphere until its intensity (or luminous flux) is reduced to 5% its original value.

1.4.5 Absorption and Attenuation

Atmospheric absorption is an important impairment to FSO communications. The atmosphere, and particularly the troposphere, consists of various gases and particles

TABLE 1.4. Spectral ranges of concern to FSO in the Atmosphere

Spectral Range	Band	Sky Transparency	Sky Brightness
1.1–1.4 microns	J	high	low at night
1.5–1.8 microns	H	high	very low

that interact with, and absorb or scatter specific wavelengths of light that enters the atmosphere. For example, water molecules absorb wavelengths above 700 nm (IR and far IR), whereas O_2 and O_3 absorb wavelengths below 300 nm (UV light).

Atmospheric opacity (or conversely, atmospheric transmittance) studies the electromagnetic radiation from the sun and from the cosmos and its selective absorption (or transmission) by the atmosphere all the way down to the ground; atmospheric opacity (or transmittance) considers the complete spectrum from radio waves to beyond gamma rays.

Most electromagnetic wavelengths are absorbed or blocked by the atmosphere; the spectrum that passes through the atmosphere and reaches the Earth's surface is known as the *optical window*. This window spans from 300 nm to about 1100 nm, and because it includes the visible spectrum 400–700 nm, it is named "optical window". Another window spans from about 2 cm to 11 m and thus it is named the "radio window", Table 1.4.

As the atmosphere absorbs light from the Sun and the cosmos, so it does to laser light that propagates through it. In fact, absorption is a critical impairment to FSO communications when there is dense fog or heavy snowfall. For example, although attenuation is a mere 0.2 dB/Km in a clear day, it can be 300 dB/Km in dense fog; in the latter case, the FSO link becomes inoperable.

In general, various molecules in the atmosphere selectively absorb wavelengths; the *absorption coefficient* for a given wavelength depends on the type of gas molecule, and on the concentration of molecules.

Water (fog, rain, snow) and liquid aerosols influence atmospheric *attenuation*. Aerosols (solid or liquid) have a very small size (from sub-nanometer to 100 nm) and thus they are suspended in the air. Consequently, each particle type in the air is responsible for light attenuation, although the actual amount of attenuation depends on molecule type, size of droplet, density of particles, and on wavelength; attenuation is measured in decibels (dB/Km). In general, the longer the λ, the lower the attenuation is, and attenuation starts becoming an issue above 5 GHz.

In general, the transmissivity of light (and from it, the absorption) is expressed by the Beer's law, also known as the Beer-Lambert law. Transmissivity of transparent matter is defined as the ratio $T = I/Io$, where I and Io are the intensity or power after crossing the matter and the incident power, respectively.

The Beer's law states that the transmissivity of light depends logarithmically on the product of the absorption coefficient, α, and the travel path of light through matter, L.

Now, if the absorption coefficient is expressed as the product of the molar absorptivity ε and the concentration c, or, as the product of an absorption cross section σ and the density N of absorbers, then, the transmissivity T is usually written as:

$$T = 10^{-aL} = 10^{-\varepsilon Lc} \text{ (for liquids)} \qquad\qquad 1.20$$

or as

$$T = e^{-\sigma LN} \text{ (for gases).} \qquad\qquad 1.21$$

Similarly the absorbance is expresses as:

$$A = -\log_{10}(I/Io) \text{ (for liquids)} \qquad\qquad 1.22$$

or as

$$A = -\ln(I/Io) \text{ (for gases).} \qquad\qquad 1.23$$

Thus, the absorbance is linear with the density of absorbers, as:

$$A = \varepsilon Lc \text{ (for liquids)} \qquad\qquad 1.24$$

or

$$A = \sigma LN \text{ (for gases).} \qquad\qquad 1.25$$

Now, because the atmosphere consists of different gases, each with different absorbance characterists, the Beer's law for the overall atmospheric absorption is:

$$I = I_0 \exp(-m(\tau_a + \tau_g + \tau_{NO_2} + \tau_w + \tau_{O_3} + \tau_r)), \qquad\qquad 1.26$$

where τ_a is the optical depth to aerosols that absorb and scatter light, τ_g is the optical depth to uniformly mixed gases (mainly carbon dioxide CO_2 and molecular oxygen O_2 that absorb light, τ_{NO2} is the optical depth to nitrogen dioxide due to atmospheric pollutants, τ_{O3} is the optical depth to ozone that absorbs light, and τ_r is the optical depth due to Rayleigh scattering; m is the *optical mass factor* (also called *airmass factor*), which is approximated to $1/\cos\theta$, where θ is the angle between the perpendicular to the Earth's surface and the observed object (the zenith angle).

Realistically, atmospheric attenuation is extremely difficult to mathematically describe, although prediction models have been devised as well as experimental or empirical models that in general are applicable to specific locations and to specific applications [36–38]. Among these models, the Longley-Rice model [39–42] predicts transmission loss for tropospheric communication and maps data over irregular terrain and it has been adopted as a standard by FCC. However, this standard pertains to signals from 20 MHz to 40 GHz and for path lengths from 1 Km to 2000 Km; thus, the optical regime is a topic that is currently under study and a reliable and efficient model is still in need. Another one is the Kruse model [43], which provides a semi-empirical formula that relates meteorological visibility to optical atmospheric attenuation, from the visible to near IR for dust and aerosol and for fog, if fog particles are much smaller than the wavelength:

$$\Gamma(V, \lambda) = (17/V) \times (550/\lambda) \times 0.581 x V^{1/3} \text{ (dB/km)} \qquad 1.27$$

where V is the visible range in Km, and λ the wavelength of the laser light. The latter formula is also simplified to:

$$\Gamma(V, \lambda) = k/V \text{ (dB/km)} \qquad 1.28$$

where k is a unit-less coefficient in the range 8.5–17 dB that depends on wavelength; Kruse predicts k = 12 at 1550 nm. If fog particles are larger than the wavelength, then the simplified Kruse formula should not be used [44].

Atmospheric emission is the opposite of absorption. When light enters the atmosphere, atoms and molecules are excited and gain energy. Excited atoms then, either spontaneously or because they were stimulated emit energy, photonic or phononic (thermal). For example, the atmosphere emits infrared radiation, which may be contained or not if there are clouds and certain gases (CO_2, H_2O) or not; when it is contained, it gives rise to the *greenhouse effect*. In addition, celestial bodies emit energy. For example, the sun at approximately 6000°K radiates electromagnetic waves at a wavelength peak of 500 nm (visible), whereas the earth at 290°K radiates at a peak of 10,000 nm (invisible).

1.4.6 Fog

Fog is a cloud in contact with the ground, it is distinguished from mist only by its droplet density, and it is expressed in degree of visibility in kilometers or in meters: fog has a higher density of droplets than mist. It reduces visibility to less than 1 Km (occasionally, to less than 50 meters), whereas mist or haze reduces visibility to no less than 2 Km. As a consequence, fog attenuates light passing through it more than mist or haze. However, the attenuation coefficient (attenuation in dB/Km) is not the same for all electromagnetic waves but it is a function of wavelength [45–47]. Therefore, in communications lower frequencies that are attenuated less by fog, are advantageous. However, such frequencies are the radio frequencies that cannot support the high bandwidth and long link length that "optical" frequencies support, which are attenuated by fog much more.

Fog begins forming when water vapor at high concentration (near 100% humidity) condenses into tiny water droplets (1–20 µm) in the air and when the difference between temperature and dew point is generally less than 2.5°C and in the presence of hydroscopic particles in the air (that stimulate vapor condensation). Water vapor is formed by the evaporation of liquid water or by the sublimation of ice (ice to vapor). The thickness of fog is largely determined by the altitude of the inversion boundary. There are a number of mechanisms that form fog, some of which are briefly described (alphabetically):

- **Advection fog** is formed when warm and moist air moves over a cool surface, such as land or ice, or even cooler waters. In such cases, the lower layers of the moist air are cooled down rapidly to form advection fog; this typically occurs during spring or fall.
- **Artificial fog** is fog generated by a water vaporizing machine and when the ambient temperature is cold. Power plants that emit large quantities of steam may also be included in this category.
- **Diamond dust** is precipitation of fine and sparse ice crystals falling from the clear sky.
- **Flash fog** is fog formed suddenly and is dissipated rapidly, as a result of temperature changing over the dew point.
- **Freezing fog** occurs when liquid fog droplets freeze to form feathery ice crystals that are deposited on the windward side of vertical surfaces, including areal wires, pylons, masts, antennas, posts, airplane wings, etc. This is common on mountain tops that are exposed to low clouds.
- **Garua fog** is a misty and transparent fog that occurs by the coast of Chile and Peru. When normal fog by the sea travels inland, it encounters hot air that causes the fog particles to shrink by evaporation and become almost invisible.
- **Ground fog** or **radiation fog** is formed when land cools after sunset by thermal radiation, in calm conditions and clear sky, when the ground quickly loses heat by radiation and cools the moist air above it to saturation point. Ground or radiation fog is localized, is dense and it occurs often in the fall and early winter.
- **Hail fog** occurs near the ground and in the vicinity of significant hail accumulation due to decreased temperature and increased moisture. This fog is localized but it can be extremely dense and abrupt.
- **Hill fog** or *upslope fog* is formed when mild moist air ascends the slope of a hill or mountain. As the air moves up the windward side of the mountain it cools down producing fog.
- **Ice fog,** aka *pogonip*, is a type of fog that consists of fine ice crystals or frozen droplets that are suspended in the air. It occurs in urban areas of Polar Regions where temperature is at or below −35°C. Ice fog can be extremely dense and may persist day and night until the temperature rises.
- **Precipitation fog**, or *frontal fog*, forms when fine rain falls into drier air below the cloud and the droplets shrink into vapor. The water vapor cools and at the dew point it condenses to form fog.
- **Steam fog** or **evaporation fog** is localized and is created by cold air passing over much warmer water or moist land.
- **Valley fog** forms in mountain valleys as a result of temperature inversion. It can last for several days in calm conditions. Valley fog is also known as Tule fog.

Because of the many parameters involved in fog characterization, empirical models have been developed [48] to characterize attenuation in terms of visibility based on

TABLE 1.5. Fog characterization based on empirical data (from [49]).

Type of Fog	Visibility (m)	Attenuation (dB/Km)
Dense fog	40–70	250–143
Thick fog	70–250	143–40
Moderate fog	250–500	40–20
Light fog	500–1000	20–9.3

data taken over a period for the location of interest [49]; as an example, the data in Table 1.5 are from [49].

1.4.7 Smog

Smog is a combined word of smoke and fog; it actually consists of air pollution. Smog is generated by factories that emit large amounts of smoke and sulfur dioxide as a result of coal burning; now, as a result of the clean air act, this type of smog has been greatly reduced in advanced countries. However, vehicular and industrial emissions interact with moisture and other molecules in the atmosphere when sunlight passes through to form secondary chemical pollutants known as partriculate matter; this process is known as *photochemical smog*. Such polutants may be aldehydes, nitrogen oxides, peroxvacyl nitrates, and other volatile organic compounds, which are reactive and oxidizing. Photochemical smog absorbs and/or scatters the laser light used in FSO communications.

1.4.8 Rain

Rain consists of water droplets in the range of 100 μm to 10 mm. As such, rain affects communications in the GHz range; for example, rain attenuation at 4 GHz is orders of magnitude less than attenuation at 12 GHz, a frequency typically used in satellite links. Rain also affects optical frequencies but not as much as snow does.

Rain attenuation of communications signals depends on droplet size, droplet density, rainfall rate, and the range of rain (or the rain cell) that the signal travels through [50]. Similarly, rain fade refers to absorption of RF signals. Notice that a signal may be influenced by rain, even though rain does not exist at the location of the transmitter or the receiver but someplace between the two.

The rainfall rate unit is mm/hr and it is measured with rain gauges over intervals of 5 minutes.

Hail is frozen rain droplets which, like rain, also affect the integrity of communication signals. However, hail has a size of 5 to 50 mm.

Like snow, rain attenuation empirical models have been developed to predict optical signal attenuation from visibility range [51–54],

$$a_{Rain} = 2.9/V \qquad\qquad\qquad 1.29$$

1.4.9 Snow

Snow, like fog and rain, has a detrimental effect in the quality of transmitted FSO optical signals.

Snow is a form of precipitation of water in the form of hexagonal ice micro-crystals due to the molecular structure of water. As these micro-crystals are created, they form flakes and come down in one of the following forms:

- **Snowflake** is a collection of snow crystals, loosely bound together into a lace structure or a puff-ball. Snowflakes can grow from 1 mm to about 10 cm across, and may be wet and sticky. A typical snowflake consists of up to 100 snow crystals.
- **Snow crystals** are single ice crystals with symmetrical shapes that grow directly from condensing water vapor in the air, usually around a nucleus of dust or some other foreign material. Snow crystals grow from microscopic up to a few millimeters in diameter.
- **Rime** is super cooled tiny water droplets (typically in a fog) that freeze quickly onto any surface, including snow crystals.

The snow attenuation exerted on optical signals, based on visibility range, is approximated by the empirical model [55]:

$$a_{snow} = 58/V \qquad\qquad\qquad 1.30$$

1.4.10 Solar Interference

The Sun's photosphere emits high intensity electromagnetic waves that are capable of interfering with electromagnetic transmission, including optical frequencies. However, solar interference occurs if the communications channel operates at the solar radiation spectrum. The solar spectrum starts from about 0.2 μm and extends beyond 2 mm, with the highest intensity in the visible range centered about 500 nm (the visible spectrum is from 400 to 700 nm) and at an average temperature that may reach ~6000°K; the Sun's radiation also includes wavelengths in the radio, Ultra Violet (UV), X-ray and Gamma-ray bands.

Optical communications transmission is typically in the range 1280 to 1620 nm. In FSO communication systems, the most popular wavelengths are about 800 nm and about 1550 nm. Thus, there is potential solar interference, although interference is more likely to be at 800 nm than 1550 nm. However, for solar interference to be troublesome, sunlight must fall onto the photodetector. Additionally, when 1550 nm is used, a long-pass optical filter at the receiver may reject wavelengths below the 1300 nm to greatly reduce the solar background radiation (SBR) interference. In FSO, the Solar Background Radiation (SBR) is greatly reduced with proper housing design and with optical filters.

1.4.11 Scattering

As light passes through the atmosphere, photons interact with molecules and other particles and are scattered in every possible direction like billiard balls; that is, photons are deflected by the molecules without altering their wavelength or energy. However, all photons do not interact the same way, because the actual scattering mechanism depends on the size and type of molecules and on the wavelength of light [56–60].

Assuming that the airborne particles in the atmosphere are small spheres with radius r, they have a refractive index n, and that light has a wavelength λ, the *size parameter* σ is defined by:

$$\sigma = \frac{4(2\pi)^5 \, r^6}{3 \, \lambda^4} \{(n^2 - 1)/n^2 + 2)\}^2 \qquad\qquad 1.31$$

Then, the intensity of scattered unpolarized photons at an angle θ, at a distance from the particle D, wavelength λ, and initial intensity I_0 is:

$$I = I_o \frac{3 \, \sigma}{16\pi D^2} (1 + \cos^2 \theta) \qquad\qquad 1.32$$

For example, nitrogen has $\sigma = 5.1 \times 10^{-31} \, m^2$ at a wavelength 532 nm, which corresponds to green light.

The above intensity equation may be modified if scattering is by molecules, which have no well-defined refractive index but they have polarizability, α.

Polarizability is defined as the ratio of the induced dipole moment P of an atom to the electric field E that produces the moment, $\alpha = P/E$; that is, polarizability is the amount of charge re-distribution of a cloud of electrons in the presence of an external electric field (of a nearby charged molecule or ion). In this case, for N scatterers, the intensity is:

$$I = I_o \frac{8 \, \pi^4 N \alpha^2}{\lambda^4 D^2} (1 + \cos^2 \theta) \qquad\qquad 1.33$$

Based on the particle size and the wavelength of light, three scattering mechanisms are identified:

- *Rayleigh Scattering* when $r \ll \lambda$; then $\sigma \sim \lambda^{-4}$
- *Mie Scattering* when $r \sim \lambda$; then $\sigma \sim \lambda^{-1.6 \text{ to } 0}$
- *Geometric Scattering* when $r \gg \lambda$; then $\sigma \sim \lambda^{\geq 0}$

Clearly, in the atmosphere where many different kinds and sizes of particles and molecules are present, scattering may occur by one or more of the above mechanisms. For example, a photon may be scattered by a very small particle (Rayleigh scattering), the scattered particle may then be scattered by a particle of the same order of magnitude

(Mie scattering) and then perhaps by a very large particle (Geometric scattering), and so on.

1.4.11.1 Rayleigh Scattering
This occurs when the particles in the atmosphere are much smaller than the wavelength of light passing through them. As a consequence, Rayleigh scattering at 400 nm is approximately 10 times greater than 700 nm for equal light intensity.

In addition, the intensity of the scattered light varies as the sixth power of the particle size and varies inversely with the fourth power of the wavelength. Because of the relatively smaller size of particles compared with the wavelength and because photons have small energy (large wavelength) and no charge, Rayleigh scattering may be studied using elastic scattering equations; that is, the energy of scattered photon before scattering and after scattering are maintained but not the direction of movement. If the photon has enough energy to excite a particle (that is, to interact with some of its vibration modes), then some energy exchange may take place; this is known as *Raman scattering*.

As a consequence of Rayleigh scattering, shorter wavelengths (blue) scatter more than longer wavelengths (red); as a result, during the day the sky is blue from all directions because what we see is the scattered light by the atmosphere's molecules and particles. When viewing the sun however, what we see is unscattered light (with blue being removed by scattering) and therefore the sun looks red-yellowish. At sunset, because the sun is at the horizon or beyond it, rays travel the longest path in the atmosphere, the blue light has already been scattered out and only the reddish rays are the ones to reach our eyes. When viewing the Earth from the outer space, the atmosphere looks blue because what one sees is scattered light, the sky looks black and the sun white. The water droplets in clouds are much larger than the wavelength of visible light, they are the result of Mie scattering, and they look white against a blue sky.

Rayleigh scattering produces forward and backward transmission patterns like antenna lobes, which are narrower and more intense for smaller particles [61].

Rayleigh scattering also occurs when light propagates in gases, liquids, as well as in transparent solid materials, such as optical fiber.

In silica fiber, the molecule density fluctuates randomly giving rise to optical power loss due to scattering. The loss coefficient due to scattering, α_{scat}, in this case is:

$$\alpha_{scat} = \frac{8\pi^3}{3\lambda^4}(n^8 p^2)(kT)\beta \qquad 1.34$$

where p is the photoelastic coefficient of silica, k is the Boltzman constant, n is the refractive index of silica, λ is the wavelength of scattered light, and β is the isothermal compressibility.

1.4.11.2 Mie Scattering
Mie scattering—the outcome of the theory of electromagnetic plane wave scattering by a dielectric sphere—occurs when the radius of the particles in the atmosphere, including aerosols are about the same size as the wave-

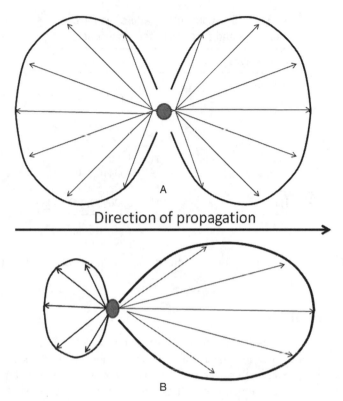

Figure 1.10. Rayleigh scattering is bidirectional (A), whereas Mie scattering is mostly unidirectional (B).

length of light passing through it [62]. Dust, pollen, smoke, and water vapor are responsible for Mie scattering. As a consequence, Mie scattering occurs in the lower layer of the troposphere where such particles are in higher concentration, particularly during overcast conditions.

Mie scattering does not strongly depend on wavelength and it produces the whitish glare around the sun when the density of particulate material is in the atmosphere, as well as the white light from mist and fog. Mie scattering also produces a transmission forward pattern like an antenna lobe, which is narrower and more intense for larger particles, Figure 1.10.

1.4.11.3 Geometric Scattering This occurs when particles in the atmosphere are many orders of magnitude larger than the wavelength; in FSO applications where laser light is 1.5 μm, the particle size is in the order of mm. Such particles may be soil, sand and dust that become airborne during storms and strong winds in specific parts of the world (deserts, agricultural areas, volcanic activity, etc). In this case, scattering and light attenuation depends on the density of particles per cubic cm or meter, and it is much more pronounced than Rayleigh or Mie scattering. As a point of

TABLE 1.6. Approximate radii of various
molecules and particles in the atmosphere.

Particle type	Radius (μm)
Air molecules	Approx. 0.0001
N2	0.000075
CO2	0.000323
O2	0.000292
Haze	0.01–1
Fog	1–20
Rain	100–10,000
Snow	1000–5000
Hail	5000–50,000

reference, Table 1.6 lists approximate radii of various molecules and particles in the atmosphere.

1.4.12 Scintillation

The atmosphere is a mixture of gases, molecules and particles that continuously gain or lose energy (heat). Some localized cells are heated more than others, whereas some others are cooled more than others. Thus, there is expansion and contraction of air cells, upward and downward movement, as well as lateral shifts. The end result is a thermal turbulence in air cells characterized by inhomogeneous and dynamically changing refractive index, density and air consistency.

When a light beam propagates through atmospheric turbulence, most of its properties are affected. That is, its polarization, refraction, absorption, scattering and attenuation fluctuate randomly at a frequency between 0.01 Hz and 200 Hz; under the same conditions, the intensity and frequency of fluctuations increase with wave frequency.

In FSO, when the laser beam crosses atmospheric turbulence, its polarization and coherency fluctuates due to anisotropic fluctuations of the air mass along the path, and its attenuation constant fluctuates due to non-consistent power loss throughout the air mass along the path [63, 64]. As a result, when the signal arrives at the receiver, its intensity fluctuates due to random temporal and spatial irradiance fluctuations of the beam, and the signal focuses and defocuses onto the photodetector randomly; this effect is similar to the flickering lights of a city in a summer night when they are viewed from far away. This signal fluctuation due to thermal turbulence is known as *scintillation*.

Scintillation affects the quality of the propagating signal and the magnitude of the *scintillation effect* is defined as the ratio between its instantaneous amplitude and its average value in the unit of time, expressed in decibel units. Several theoretical models have been developed in an attempt to predict the scintillation effect [65, 66]. However, scintillation is a complex phenomenon that also depends on time-variant meteorological phenomena, which are difficult to model. Therefore, theoretical scintillation models are valuable as long as the parameters and boundary conditions they were developed for hold, and they also require validation by experimental data.

Scintillations is measured by the scintillation index, σ_{index}, which is expressed as:

$$\sigma_{index}^2 = (< I^2 > - < I >^2)/ < I >^2 \qquad 1.35$$

where I is the irradiance (or intensity) of the optical wave, and the brackets $< >$ denote the average value over time.

In FSO communication links as well as in laser radar links, knowledge of the scintillation index is important for determining system performance. The scintillation index increases with path length, while the optical signal becomes less coherent and the focusing effect at the optical power receiver deteriorates.

To ameliorate the scintillation effect, the diameter of the collecting lens at the receiver is increased to beyond the *irradiance correlation width*, ρ_c, of the received optical signal; the irradiance correlation width is determined from the irradiance covariance function, which is a Gaussian function of time, and it identifies the maximum receiver aperture size that acts like a *point receiver*.

Apertures larger than ρ_c perform *aperture averaging*, and as such, the *correlation width* ρ_c is a design parameter that helps to reduce the amount of scintillation effect as is experienced by the photodetector of the receiver. An interesting method that uses a rotation pipe to reduce the amount of speckle has been reported [67].

1.4.13 Wind and Beam Wander

Wind is movement of air masses that takes place in the troposphere and in the stratosphere [68] Light itself is not affected by wind only, but perhaps what causes wind (temperature variation and other factors). However, in FSO applications, wind has a degrading effect on beam alignment because FSO transceivers are positioned on tall buildings or on poles. Strong wind turbulence can sway a tall building by few meters and thus displace the beam from its target receiver. As a consequence, the effect of turbulent winds on beam alignment may be significant as the laser beam seems to wander at the receiver; this is known as *beam wander*. If in addition to wind, there are localized temperature variations on the beam path (which cause scintillation), then beam wander may also affect the scintillation index [69]; this is known as *beam wander induced scintillation* (BWIS). Clearly, the effect of the latter depends on beam properties; BWIS is insignificant if the beam is collimated or is divergent, and is aggravated if the beam is focused. Currently, we have no significant data to identify the effect on BWIS of a very thin, almost not divergent, laser beam with and without auto-tracking.

In short, the effect of beam wander depends on whether the beam is collimated, is divergent or is focused, on the diameter of the beam, and also whether the transceivers have an automatic tracking system to maintain alignment even if the position of one transceiver shifts by few meters with respect to the other.

1.5 CODING FOR ATMOSPHERIC OPTICAL PROPAGATION

FSO communications is based on an optical narrow beam that propagates through the atmosphere. The atmosphere, an unbounded medium, is very dynamic and its parameters are not constant. Therefore, as already described, the various atmospheric phenomena of the troposphere have an adverse effect on the quality of the optical signal. Therefore, the obvious question: which modulation method of the optical signal would be more effective?

Drawing from the experience in fiber optic communications, currently four methods seem to be the most probable: phase-shift keying (PSK), frequency shift keying (FSK), state of polarization shift keying (SoPSK), and amplitude or on-off keying (OOK).

- The PSK is effective if the phase remains unadulterated in the medium and along the transmission path. As we have discussed, the phase of the FSO beam changes due to refractive index changes and scintillation. Therefore, it is obvious that this modulation method will produce an increased bit error rate (BER).
- The FSK requires switching between two different optical frequencies and thus it requires two laser transmitters, or a laser transmitter and a wavelength converter device. Although this method could be very effective, it adds to the cost and to the design complexity of the transceivers.
- The SoPSK experiences the same issues with the PSK; the polarization is not maintained by the air medium due to the dynamic changes in refractive index of the air.
- The OOK modulation method is the most basic form of pulsed modulation and is used in binary direct detection receivers in fiber-optic and in FSO communications, for which the link performance versus link length, the signal to noise ratio (SNR), and the bit error rate (BER) have been thoroughly studied.

The performance (probability of error in terms of SNR) for OOK is:

$$P_e = 1/2 \text{ erfc } \sqrt{(S/N)} \qquad\qquad 1.36$$

where *erfc* is the error function complimentary.

In FSO systems, the OOK method seems to be the most robust of the four aforementioned since pulsed light with direct detection does not depend as much on phase and polarization variations, although it does on intensity.

In addition, because the performance of an FSO link can be deduced from the probability density function (PDF) of the irradiance signal, then, in the presence of air turbulence the unconditional BER is determined from the probability of error or the conditional probability averaged over the PDF of the random signal.

1.6 LIDAR

LIDAR is an acronym that stands for *Laser Detection And Ranging*. As the acronym states, it is based on a thin laser beam that is pulsed and emitted in the atmosphere; the

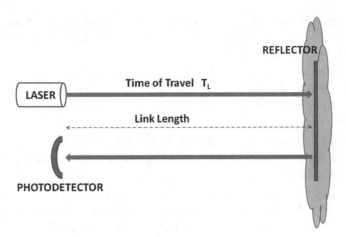

Figure 1.11. Principles of a LIDAR.

power of the beam may be from few milliWatts to many Watts. In addition, a photo-detector measures the beam reflection to determine phase shift, intensity, and perhaps polarization state, Figure 1.11. The laser beam may be reflected vertically or horizontally, and it can be stationary or moving in a rotational mode or in a scan mode. As such, LIDARs have found applications in astronomy, in meteorology, in archaeology, in land surveying, in 3-D object mapping, distance measurement, object detection, movement detection, and more [70–72].

LIDARs use a short wavelength that typically ranges from the UV (below 400 nm) to IR (above 700 nm); that is, the complete visible spectrum and beyond. As such, the feature accuracy of objects obtained by LIDARs are very accurate, in the order of the wavelength. However, the laser beam is also susceptible to atmospheric impairments (absorption, attenuation, scattering, scintillation, etc.) as a result of molecular, aerosol, and atmospheric pollutant interaction. Although this may seem to be detrimental to LIDAR applications, in fact it is an advantage: because of the small wavelength, reflections take place by minute dielectric discontinuities or localized dielectric anomalies (caused by aerosols, smoke, and other pollutants) and not by large metallic objects as in the case of radio frequency RADARs; the long wavelength of RADAR is insensitive to small particles, like large sea waves over small pebbles are. Thus, the LIDAR has been used in atmospheric research, in weather prediction, and in air pollution identification and measuring.

Depending on the type and size of molecules that scatter the beam, LIDARs are classified in:

- Rayleigh LiDAR,
- Mie LiDAR,
- Raman LiDAR, which exploits inelastic scattering; a small amount of the photonic power interacts with gases and light is scattered with a longer wavelength which depends on the type of gas. This LIDAR is used for atmospheric gas

concentration, and for aerosol parameter measurement), as already described. Using THz interferometry, chemical, nuclear, or organic molecules can be detected from a distance.

- The differential absorption LIDAR uses an "on-line" or probe wavelength that is absorbed by the gas and another wavelength that is not absorbed. From the scattered light, the differential absorption is measured from which ozone, carbon dioxide, or water vapor measurements are made.
- The Elastic Backscatter (by aerosols and clouds) LIDAR.
- The Fluorescence LiDAR is based on the fluoresce effect of certain elements (Na, Fe, K), and
- The Doppler LIDAR is similar to the Elastic Backscatter LIDAR but it measures the frequency shift of the backscattered light from which the wind speed is calculated. Traffic speed enforcement uses this type of LIDAR.

Moreover, there are two more orthogonal classification types of LIDARs, the coherent light and the incoherent type.

- The coherent type LIDAR uses coherent light and measurements are based on phase differences of the reflected beam. Thus, the receiver must be a heterodyne photodetector receiver. This type is suitable for moving objects and for phase sensitive measurements.
- The incoherent type LIDAR uses incoherent light and measurements are based on the amplitude (intensity) of the reflected beam. That is, the receiver uses a simple direct intensity photodetector.

When the LIDAR beam scans the surface of a 3-D object, the beam is reflected and a map of its surface may be obtained by measuring the time and/or the phase of the pulsed light that returns to the detector; thus the distance is measured with resolution commensurate with the wavelength (nm) or the pulse repetition rate. The information obtained may be used to identify an object, to map or survey a terrain, to reproduce (copy) an object, and more. One of the simplest LIDAR applications is the label and barcode scanner at check-out points.

Regarding laser power, in atmospheric applications it is many Watts whereas in surveying and identifying objects is few milliWatts. More LIDAR applications are emerging including automatic vehicular guidance and control, and possibly automatic control of FSO link alignment and performance.

REFERENCES

1. S.V. Kartalopoulos, "A Global Multi Satellite Network", U.S. patent #5,602,838, 2/11/1997.
2. K. Shaik, "Atmosphere Propagation Effects Relevant to Optical Communications," TDA Progress Report, 42–94, pp. 180–200, Jet Propulsion Laboratory, Pasadena, CA, August, 1988.

3. S.V. Kartalopoulos, *DWDM: Networks, Devices and Technology*, IEEE/Wiley, 2003.

4. S.V. Kartalopoulos, *Introduction to DWDM Technology: Data in a Rainbow*, IEEE/Wiley, 2000.

5. ITU-T Recommendation G.652, *"Characteristics of a single-mode optical fibre cable"*, Oct. 2000 (Table G.652.C lists the parameters of the water-free fiber).

6. ITU-T Recommendation G.692, "Optical interfaces for Multi-channel Systems with Optical Amplifiers", Oct. 1998 (Appendix VII provides bidirectional WDM transmission recommendations, Appendix VIII transmission of 16 and 32 channels), and Corrigentum 1, Jan. 2000.

7. ITU-T Recommendation G.694.1, "Spectral Grids for WDM Applications: DWDM Frequency Grid", 5/2002.

8. ITU-T Recommendation G.694.2, "Spectral Grids for WDM Applications: CWDM Wavelength Grid", 6/2002 Draft.

9. B.E.A. Saleh and M.C. Teich, *Fundamentals of Photonics*, John Wiley & Sons, New York, 1991.

10. F. Pampaloni and J. Enderlein (2004). "Gaussian, Hermite-Gaussian, and Laguerre-Gaussian beams: A primer". *ArXiv:physics/0410021*. http://arxiv.org/abs/physics/0410021.

11. A Tutorial on Gaussian Beam Optics, Newport: http://www.newport.com/servicesupport/Tutorials/default.aspx?id=122; Retrieved Jan 2011.

12. L. Andrews, R.L. Philips, and C.Y. Hopen, *Laser Beam Scintillation with Applications*, SPIE Press, 2001.

13. ANSI Z136.1-1993, "American National Standard for Safe Use of Lasers".

14. *U.S. Standard Atmosphere*, U.S. Government Printing Office, Washington, D.C., 1976. Also in http://ntrs.nasa.gov/archive/nasa/casi.ntrs.nasa.gov/19770009539_1977009539.pdf.

15. F.K. Lutgens and E.J. Tarbuck, *The Atmosphere*, Prentice Hall, 1995.

16. ITU-R Recommendation P.530-11, "Propagation data and prediction methods required for the design of terrestrial line-of-sight systems," International Telecommunication Union, Geneva, 2005.

17. B.A. Bodhaine, N.B. Wood, E.G. Dutton, and J.R. Slusser, "On Rayleigh Optical Depth Calculations," *Journal of Atmospheric and Oceanic Technology*, vol. 16, issue 11, pp. 1854–1861, Nov. 1999. Good historical overview of refractive index modeling.

18. H. Barrell and J.E. Sears, "The Refraction and Dispersion of Air for the Visible Spectrum," *Philosophical Tranactions of the Royal Society A*, vol. 238, pp. 6–62, 1939.

19. B. Edlén, "Dispersion of standard air," *Journal of the Optical Society of America*, vol. 43, pp. 339–344, 1953.

20. B. Edlén, "The refractive index of air," *Metrologia*, vol. 2, pp. 71–80, 1966.

21. J.C. Owens, "Optical refractive index of air: dependence on pressure, temperature and composition," *Applied Optics*, vol. 6, pp. 51–59, 1967.

22. E.R. Peck and K. Reeder, "Dispersion of air," *Journal of the Optical Society of America*, vol. 62, pp. 958–962, 1972.

23. J.W. Marini and C.W. Murray, Jr, *Correction of laser range tracking data for atmospheric refraction at elevations above 10 degrees*, NASA-TM-X-70555, 1973.

24. C.S. Gardner, *Correction of laser tracking data for the effects of horizontal refractivity gradients*, Applied Optics, vol. 16, pp. 2427–2432, 1977.

25. F.E. Jones, "The refractivity of air," *Journal Research NBS*, vol. 86, pp. 27–32, 1981.

26. P.E. Ciddor, "Refractive index of air: new equations for the visible and near infrared," *Applied Optics*, vol. 35, pp. 1566–1573, 1996.

27. P.E. Ciddor and R.J. Hill, "Refractive index of air," *Applied Optics*, vol. 38, pp. 1663–1667, 1999.

28. V. Mendes, *Modeling the neutral-atmosphere propagation delay in radiometric space techniques*, PhD. thesis, University of New Brunswick, 1999.

29. H. Yan and G. Wang, "New consideration of atmospheric refraction in laser ranging data," *Monthly Notices of the Royal Astronomical Society*, vol. 307, pp. 605–610, 1999.

30. Y.S. Galkin, R. Tatevian, and L. Blank, *Correction of the water vapour absorption line effect for EDM with infrared emitting diodes*, 22nd General Assembly of the International Union of Geodesy and Geophysics (IUGG), 1999.

31. P.E. Ciddor, "Refractive Index of Air: 3. The Roles of CO_2, H_2O, and Refractivity Virials", *Applied Optics*, vol. 41, pp. 2292–2298, 2002.

32. V.A. Rakov and M.A. Uman, *Lightning Physics and Effects*, Cambridge Univ. Press, 2003.

33. T.F. Malone, Ed, *Compendium of Meteorology*, American Meteorological Society, Boston, Mass, 1951.

34. N. Never, *Air Pollution Control Engineering*, McGraw-HILL, Singapore, 1995.

35. E.R. Cohen and B.N. Taylor, *Journal of Research National Bureau of Standards*, vol. 92, pp. 85–95, 1987. (International Union of Pure and Applied Chemistry (IUPAC))

36. ITU-R Recommendation P.618-7, 2001, "Propagation Data and Prediction Methods Required for the Design of Earth-Space Telecommunication Systems".

37. H. Hemmati, *Deep Space Optical Communications*, John Wiley & Sons, 2006.

38. P.W. Kruse and al., "Elements of infrared technology: Generation, transmission and detection", J. Wiley and sons, New York, 1962.

39. A.G. Longley, "Radio propagation in urban areas," OT Report 78–144, Apr. 1978.

40. A.G. Longley, "Local variability of transmission loss-land mobile and broadcast systems", OT Report, May 1976.

41. OET BULLETIN No. 69, *Longley-Rice Methodology for Evaluating TV Coverage and Interference*, February 2006; it provides guidance on the implementation and use of Longley-Rice methodology.

42. http://www.v-soft.com/probe/probeIIgalary.htm provides a gallery of map studies using the Longley-Rice model. Retrieved Nov 23, 2010.

43. P.W. Kruse, L. McGlauchlin, and O.H. Vaughan, *Elements of Infrared Technology: Generation, Transmission and Detection*, John Wiley & Sons, New York, 1962.

44. R.M. Pierce, J. Ramaprasad, and E. Eisenberg, "Optical Attenuation in Fog and Clouds," *Optical Wireless Communications IV, Proceedings of SPIE*, vol. 4530, pp. 58–71, 2001.

45. I.I. Kim, B. McArthur, and E. Korevaar, "Comparison of laser beam propagation at 785 nm and 1550 nm in fog and haze for optical wireless communications," *Proc. SPIE*, 4214, 26–37, 2001.

46. M. Al Naboulsi, H. Sizun, and F. de Fornel, Fog Attenuation Prediction for Optical and Infrared Waves, Journal SPIE, International Society for Optical Engineering, 2003.

47. M. Gebbart, E. Leitgeb, M. Al Naboulsi, H. Sizun, and F. de Fornel, Measurements of light attenuation at different wavelengths in dense fog conditions for FSO applications, STSM-7, COST270, 2004.

48. S.S. Muhammad, B. Flecker, E. Leitgeb, and M. Gebhart, "Characterization of fog attenuation in terrestrial free space optical links," *Journal of Optical Engineering*, vol. 46, no. 4. Paper. 066001, June 2007.

49. M.S. Awan, L.C. Horwath, S.S. Muhammad, E. Leitgeb, F. Nadeem, and M.S. Khan, "Characterization of Fog and Snow Attenuations for Free-Space Optical Propagation," *Journal of Communications*, vol. 4, no. 8, pp. 533–545, September 2009.

50. M. Akiba, K. Wakamori, and S. Ito, *"Measurement of optical propagation characteristics for free-space optical communications during rainfall"*, IEICE Transactions on Communications E87-B, 2053–2056 (2004).

51. D. Atlas, "Shorter Contribution Optical Extinction by Rainfall," *J. Meteorology*, vol. 10, pp. 486–488, 1953.

52. ITU-R Recommendation P.839-3, "Rain height model for prediction models", International Telecommunication Union, Geneva, 2001.

53. ITU-R, "Development towards a model for combined rain and sleet attenuation", ITUR Document 3M/62E, International Telecommunication Union, Geneva, 2002.

54. ITU-R Recommendation P.837-4, "Characteristics of precipitation for propagation modeling", International Telecommunication Union, Geneva, 2003.

55. H.W. O'Brien, "Visibility and Light Attenuation in Falling Snow," *Journal of Applied Meteorology*, vol. 9, pp. 671–683, 1970.

56. H.C. van de Hulst, *Light scattering by small particles*, New York, Dover, 1981.

57. M. Kerker, *The scattering of light and other electromagnetic radiation*, Academic Press, New York, 1969.

58. C.F. Bohren and D.R. Huffmann, *Absorption and scattering of light by small particles*, John Wiley-Interscience, New York, 1983.

59. P.W. Barber and S.S. Hill, *Light scattering by particles: Computational methods*, World Scientific, Singapore, 1990.

60. D. Atlas, M. Kerker, and W. Hitschfeld, "Scattering and attenuation by nonspherical atmospheric particles," *Journal for Atmospheric and Terrestrial Physics*, vol. 3, pp. 108–119, 1953.

61. M. Sneep and W. Ubachs, "Direct measurement of the Rayleigh scattering cross section in various gases," *Journal of Quantitative Spectroscopy and Radiative Transfer*, vol. 92, p. 293, 2005.

62. G. Mie, "Beiträge zur Optik trüber Medien, speziell kolloidaler Metallösungen," *Leipzig, Ann. Phys*, vol. 330, pp. 377–445, 1908.

63. L.C. Andrews and R.L. Phillips, *Laser Beam Propagation through Random Media*, 2nd ed., SPIE SPIE Press, 2005.

64. *Encyclopedia of Optical Engineering*, Volume 3, R.G. Driggers, Editor, CRC Press, 2003.

65. J. Sala, M. Lamarca, J. A. Lopez, F. Rey, J. Riba, G. Vazquez, X. Villares, A. M. Jalon, and P. Rodrıguez, "A Rain and Scintillation Ka-band Channel Simulator", 10th International Workshop on Signal Processing for Space Communications (SPSC 2008), 6–8 October 2008, Rhodes Island, Greece. Paper also available at: http://www.gts.tsc.uvigo.es/gpsc/sproactive/Documents/Papers/SPSC08.pdf.

66. L.C. Andrews, R.L. Phillips, and C.Y. Hopen, *Laser Beam Scintillation with Applications*, SPIE Press, 2001.

67. M. Sun and Z. Lu, "Speckle suppression with a rotating light pipe," *Optical Engineering*, vol. 49, no. 2, paper 024202, February 2010.

68. J.A. Dutton, *The Ceaseless Wind: An Introduction to the Theory of Atmospheric Motion*, Dover Publications, New York, 1986.

69. L.C. Andrews, et al., "Beam wander effects on the scintillation index of a focused beam," *SPIE*, vol. 5793, 2005.

70. http://home.iitk.ac.in/~blohani/LiDAR_Tutorial/Airborne_AltimetricLidar_Tutorial.htm provides a quick tutorial of LIDARs and their applications. Retrieved Nov. 23, 2010.

71. http://www-calipso.larc.nasa.gov/ provides a description of the Cloud-Aerosol Lidar and Infrared Pathfinder Satellite Observation (CALIPSO) satellite used for measurements of clouds and atmospheric aerosols (airborne particles) that play a role in regulating Earth's weather, climate, and air quality.

72. http://ramanlidar.gsfc.nasa.gov/ describes NASA's Raman LIDAR for measuring water vapor, aerosols and other atmospheric species.

2

FSO TRANSCEIVER DESIGN

2.1 INTRODUCTION

Free-space optical (FSO) is a line-of-sight technology that uses modulated laser light in the unlicensed spectrum to transmit multiprotocol data (voice, video, music, and data) through the atmosphere. FSO is formidable to applications that require to establish an operational link semi-permanently or quickly, within a day or so, and where a fiber infrastructure is not in place; such applications pertain to enterprise, to short term communication needs, to disaster areas, to areas where fiber installation is not cost-effective, and to areas where severe atmospheric phenomena (dense fog, smog and snow) are not the norm.

Currently, the primary focus on FSO research and development is toward higher data rate, and longer links in a mesh network topology; among the technological challenges to address are sun glare minimization, operation in adverse temperature limits, accurate laser beam pointing, alignment management between multiple transceivers per node and alignment maintaining (known as auto-tracking function), security (data and network), and communications protocol.

In this chapter, we focus on the technology of components that comprise the transceiver, and we address issues associated with node design.

Free Space Optical Networks for Ultra-Broad Band Services, First Edition. Stamatios V. Kartalopoulos.
© 2011 Institute of Electrical and Electronics Engineers. Published 2011 by John Wiley & Sons, Inc.

2.2 LIGHT SOURCES

2.2.1 Laser Classification Based on Reach

In Chapter 1 (1.2.13), the laser classifications from a safety viewpoint, old and revised, were listed. In communications, lasers also are classified based on reach, which is the maximum distance at optical power sufficient to be detected. Clearly, for a given fiber, the longer the reach the more powerful the laser should be. Thus, in communications, the laser classification based on reach is:

- *Extended Long Reach* if the driving optical power is up to 80 Km SMF (Lasers),
- *Long Reach* if the driving optical power is up to 40 Km on SMF (Lasers),
- *Intermediate Reach* if the driving optical power is up to 15 Km on MMF or SMF (Lasers),
- *Short Reach* if the driving optical power is up to 2 Km on MMF,
- *Very Short Reach* if the driving optical power is less than 1 Km on glass MMF or even plastic fiber (LEDs)

Exercise 1: A Short Reach laser from a multimode fiber (MMF) application is used in an FSO application. The attenuation at 1310 nm is 1 dB/Km in MMF. Estimate the maximum FSO link length if the atmospheric attenuation in a foggy day is 20 dB/Km. Assume all other factors unimportant.

Solution 1: The Short Reach laser is defined at less than 2 Km in MMF. The attenuation constant in MMF is 1 dB/Km. If the fog has 20 dB/Km, then the laser beam transmitted through fog will be 2/20 Km = 0.1 Km.

Exercise 2: If in Exercise 1, one considers a divergent beam, then, does this affect the calculated link length through fog?

Solution 2: Because the power density of the divergent beam decreases exponentially with the distance from the source, as the receiver moves closer to the source, the power density also increases exponentially. Therefore, the actual link length is greater than 0.1 Km.

2.2.2 Parameters of Laser Sources

ITU-T Recommendation G. 650 defines parameters for many optical components. Here we add a list of parameters pertinent to light sources. Some of these parameters have been described in Chapter 1, and the remaining will be described in subsequent sections.

- Optical output power, P_o
- Optical center wavelength, λ_0
- Channel spacing, $\Delta\lambda$
- Cut-off wavelength (tunable sources)
- Spectral width (tunable sources)

- Line width, $\delta\lambda$
- Beam profile: cross-section distribution and modes
- Beam divergence
- Modulation depth (modulated sources)
- Bit rate (min-max; modulated sources)
- Source noise
- Source chirp
- Wavelength and power dependency on bias
- Wavelength and power dependency on temperature

2.2.3 Light Emitting Diodes

A light-emitting-diode (LED) is a monolithically integrated p-n semiconductor device, that with characteristics similar to the common semiconductor diode. When the p-n junction is forward biased, during the electron-hole recombination process at the junction, energy is released; those with sufficient energy emit energy in the form of light (photons), and those with insufficient energy generate energy in the form of heat. The amount of photons generated depends on the amount of forward bias.

As the recombination is a statistical random event, the LED light is not coherent. Optical power emerges from the device edge in a relatively large cone and its amplitude depends on the current density, which also depends on the electron concentration and the applied voltage. Electrically, LEDs exhibit the same I-V characteristics of common diodes. Moreover, a threshold is defined, below which the optical power is negligible.

The *switching speed* of LEDs depends on the recombination rate, R, and is expressed by:

$$R = J/(de) \qquad\qquad 2.1$$

where J is the current density (A/m^2), d is the thickness of the recombination region, and e is the electron charge.

The *output power* of LEDs is expressed by:

$$P_{out} = \{(\eta hc)/(e\lambda)\}I \qquad\qquad 2.2$$

where I is the LED drive current (A), η is the quantum efficiency (relative recombination/ total recombination), h is Plank's constant, e is the electron charge and λ is the wavelength of light.

The output optical spectrum of LEDs is the range of emitted wavelengths. This depends on the absolute junction temperature (i.e., the range widens as temperature increases) and on the emission wavelength λ:

$$\Delta\lambda = 3.3(kT/h)(\lambda^2/c) \qquad\qquad 2.3$$

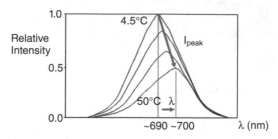

Figure 2.1. The efficiency and the accuracy of a laser depends on temperature.

where T is the absolute temperature at the junction, c is the speed of light, k is Boltzmann's constant, h is Plank's constant.

Temperature has an adverse effect on the stability of an LED device. As temperature rises, its wavelength shifts and its intensity decreases, Figure 2.1.

Intensity decrease reduces the signal to noise ratio and the ability to transmit the signal to long fiber length, whereas wavelength shift may have an unpleasant effect on cross-talk and bit error rate increase.

The modulated current density J is expressed by:

$$J = J_o + J_o m_j \exp(j\omega t) \qquad\qquad 2.4$$

where J_0 is the steady state current density, m_j is the modulation depth, and ω is the modulation frequency. This current modulates the electron density difference through the junction, $\Delta n = n - n_o$ (n_o is the electron density at equilibrium with no bias current) as:

$$\Delta n = N_o\{1 + M_N \exp[j(\omega t - \theta)]\} \qquad\qquad 2.5$$

where N_0 is the electron density at steady state, M_N is the electron modulation depth, and θ is the phase shift. From the differential $d(\Delta n)/dt$ the output power modulation index, I_M, is derived in terms of the output modulation response, M_N:

$$I_M = M_N \exp(-j\theta) = m_j/(1 + j\omega\tau_r) \qquad\qquad 2.6$$

where τ_r is the electron-hole recombination time.

Comparing the modulation response with a first-order low-pass (LP) filter, it is concluded that their transfer functions are identical. Thus, the modulation response may be studied like an LP filter, from which the 3 dB modulation bandwidth is derived:

$$\omega_{3db} = 1/\tau_r \qquad\qquad 2.7$$

Based on the above analysis, the salient features of LEDs and their applicability are summarized as follows:

- Their bandwidth depends on device material
- Their optical power depends on current density (that is, on the operating V-I point)
- Optical power and spectrum depends on temperature
- The emitted light is not coherent
- They are relatively slow devices (<1 Gb/s); in communications they may be used only in low bit-rates.
- They transmit light in a relatively wide cone; in communications they may be used only over MMF.
- They exhibit a relatively wide spectral range
- They are inexpensive
- LEDs have been created to emit red, green, blue and white light and they have found many applications including automotive and residential; because of their lower power consumption and their increased emitted power, their applicability keeps increasing. In fact, combined with phosphorescent materials, LEDs have replaced the incandescent light-bulb in many applications.

2.2.4 LASERs

LASER is an acronym that stands for light amplification by stimulated emission of radiation. They are devices that contain some elements, which when in gaseous state (e.g. He-Ne) or doped in crystals (e.g. ruby with 0.05% Chromium) absorb electrical or electromagnetic energy and they transit in a semi-stable excited energy state. When in excited state, other photons of a specific wavelength that travel through the material stimulate the excited atoms, which release photonic and or acoustic energy and they transit to a lower energy level. The actual process of excitation and stimulation varies from element to element and the excitation and stimulation mechanism may be simple or complex. It all depends on the quantum energy levels of particular electrons/atoms, the energy available to excite them, and the energy required to stimulate them. Figure 2.2 illustrates three different transition systems, a 3-energy-level, a 2-energy, and a multi-energy-level.

In lasers, stimulated photons enter a region known as the resonant cavity in which a strong directional and *coherent* monochromatic beam is formed; photons traveling in other directions are lost through the walls of the cavity and do not contribute to the laser beam.

The resonant cavity has specific dimensions with polished ends or gratings to form reflectors forming a frequency-selective mechanism that produces an optical beam within a narrow spectrum. As energy is pumped in and the excitation and stimulation process continues, the optical gain reaches a threshold and the lasing process starts. Thus, depending on semiconductor composition, structure, pumped energy and feedback mechanism, lasers have a large positive gain.

Like all resonant cavities, so the laser cavity may support a number of frequencies (wavelengths) that meet the condition $\lambda = (2 \times L)/N$, where N is an integer and L is

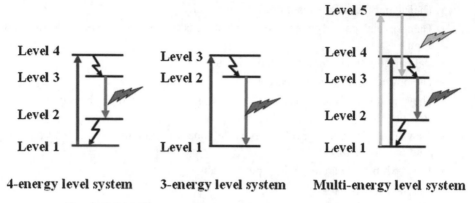

Figure 2.2. Different excitation and stimulation transition systems.

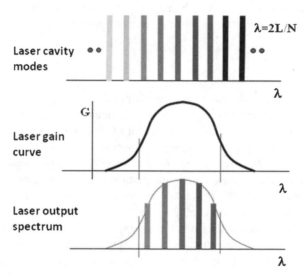

Figure 2.3. The frequencies supported depend on the specific laser.

the cavity length. However, the upper and lower bound of wavelengths as well as the amplitude of each wavelength depend on the gain bandwidth of the laser, Figure 2.3.

Ideally, the generated beam should have an intensity symmetrically distributed, as discussed in Chapter 1. However, the resonant cavity determines the actual cross-section distribution of the emerging beam, or transverse mode, which can deviate from ideal yielding more complex cross-sections.

Semiconductor materials such as AlGaAs and InGaAsP generate photons with wavelengths compatible with the low loss regions of silica fiber and compatible with other optical components. The active layer of this structure, such as a straight channel of InGaAsP, is sandwiched between n- and p-type layers of InP, also known as *cladding layers*. When bias is applied, the recombination of holes and electrons in the active region releases light, the wavelength of which depends on the energy band-gap of the active material. The active layer has a much higher refractive index than the cladding layers and thus the cladding layers confine the electron-hole pairs and the photons in the active region. The active region forms a resonant cavity that supports coherent photons of a selected optical frequency (wavelength). These photons are coherent (because of the resonant cavity) and form a beam within a very narrow cone. In many lasers, the generated beam is guided so that about 95% is emitted at the front-face of the device and the remainder at the back-face for monitoring purposes. The ratio front-to-back of output power is known as the *tracking ratio*.

The laser beam generated in the active area of the laser device may be directly or indirectly modulated. Direct modulation at very high rates (10 Gb/s and higher) may cause lasers to optically chirp. Optical chirping is observed as a spectral line that jitters about the central wavelength. Optical chirping occurs because the refractive index of the laser cavity depends on the drive current. As the drive current changes abruptly from logic ONE to logic ZERO and vice versa, the refractive index changes dynamically and thus the resonant cavity characteristics; this causes a dynamic change in wavelength, which dynamically broadens the laser line width, and hence *optical chirping*. Chirping is avoided if external modulation is used in which case the laser emits a *continuous wave* (CW). When lasers and modulators (made with In+Ga+As+P) are monolithically integrated on InP substrate, electrical isolation is required to minimize chirping. At low bit rates, chirping is easier tolerated.

Device compactness is very important to communications and currently several key optical functions, such as lasers, filters, modulators, multiplexers, and others, are currently integrated using advanced monolithic methods to produce components with more functionality per cube unit that operate efficiently over a wide temperature range.

Wavelength and signal amplitude stability of semiconductor lasers is important in all applications. Stability depends on material, bias voltage and temperature. In high-bit-rate applications, frequency and amplitude stabilization is achieved using thermo-electric cooling devices that can keep the temperature stable within a fraction of a degree Celsius. However, this adds to cost and power consumption and "cooler" devices at higher optical powers and bit rates are in development.

Fixed wavelength CW lasers used in communications have a typical output power in the range 10–30 mW, linewidth better than 10 Mhz, single mode suppression ratio (SMSR) less than 50 dB, and wavelength stability 10 pm (+/− 1.25 Ghz). Moreover, lasers may support a *single transverse mode* (known as *single-mode lasers*), or both a *single transverse mode* and a *single longitudinal mode* (*single-frequency lasers*), or they may oscillate at *several frequencies simultaneously* (*multi-frequency lasers*). Finally, lasers may have a single *fixed frequency* or they may be *tunable*.

2.2.4.1 Fabry-Perot Semiconductor Lasers
Fabry-Perot semiconductor lasers are based on the Fabry-Perot (F-B) cavity resonator principle. The F-B Laser consists of a semiconductor material in the form of a straight channel (p-type AlGaAs),

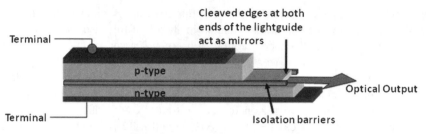

Figure 2.4. Simplified F-P semiconductor laser

which is both active (for excitation and stimulation) and optical waveguide (to guide photons in one direction). A simplified F-B laser structure is shown in Figure 2.4.

Both ends of the channel are cleaved to act as semi-mirrors, with reflectivity.

$$R = \{(n-1)/(n+1)\}^2 \qquad\qquad 2.8$$

where n is the refractive index of the active medium.

Because a Fabry-Perot resonator supports multiple wavelengths [1], so does the F-P Laser. Thus, F-B Lasers generate several longitudinal frequencies (modes) at the same time. The semiconductor laser material, the frequency spacing and the F-B Laser determine the range of frequencies, whereas the bias current determines the threshold frequency.

If the two reflectors of the Fabry-Perot resonator are external to the active region, then, changes in the geometry between the two mirrors change the resonance characteristics of the cavity, and the produced wavelength achieving tunability.

2.2.4.2 Bragg Lasers Inaccurate reflectivity and flatness of the semi-mirror edges of the Fabry-Perot resonators may result in laser light that lacks the required spectral quality. Employing Bragg gratings as reflectors, a narrower spectrum is achieved. Bragg gratings at either side of the active area are achieved by periodically varying the index of refraction of the material. However, because Bragg gratings act as full reflectors, an optical waveguide is placed right under the active area and the generated light is coupled onto it from which is emerges from the Laser device, Figure 2.5. Such lasers are known as distributed Bragg reflectors (DBR).

Distributed Feedback (DFB) lasers are monolithic devices with an internal structure based on InGaAsP waveguide and internal grating technologies, typically at the interface n-InP substrate and n-InGaAsP layers, to generate light at a fixed wavelength determined by the active area and the grating. The DFB structure may be combined with multiple quantum well (MQW) structures to improve the linewidth of the produced laser light (make it as narrow as few hundred KHz). MQWs have a similar structure with the diode structure but the active junction is few atomic layers thin—see next section).

2.2.4.3 VCSEL Lasers Fabry-Perot and Bragg lasers require electrical current in the order of tens of milliamperes. Moreover, their output beam has an elliptical

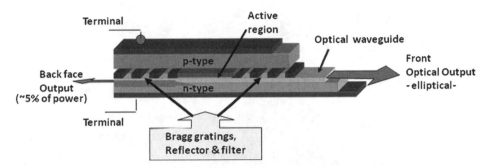

Figure 2.5. Two Bragg gratings bound the laser active area acting as reflectors.

cross-section, typically at an aspect ratio of 3:1. Fortunately, added passive optics such as a half-cylinder microlens, can convert an elliptical beam to cylindrical.

Semiconductor quantum well lasers (QWL) are lasers, the active junction of which is 50 to 100 Angstroms or 5 to 10 atomic layers thin. Very thin layers are grown using molecular beam epitaxy (MBE) or metal organic chemical vapor deposition (MOCVD).

The active region consists of a GaAs quantum well layer sandwiched between a p-type $Al_xGa_{1-x}As$ layer and n-type $Al_yGa_{1-y}As$ layer. In principle, a stack of many p- and n-type AlGaAs and quantum-well GaAs layers are grown on a thick n-type GaAs substrate, topped by a last p-type GaAs thick layer. This structure is finally sandwiched between two metallic electrodes (made of Au/Zn) for the bias voltage and for letting photons transmit through; the Au/Zn electrode can be made semitransparent or totally reflective. The p-type and n-type layers are made so that they comprise Bragg reflectors. More complex structures with multiple quantum wells produce what is termed *multiple quantum well lasers* (MQWL or MQW). Other structures are based on indium phosphide [2–4].

Quantum well lasers have a very interesting property. A bias current excites the active region and generates electron-holes which when recombined produce photons. The generated electron-holes are confined to move in an almost horizontal plane, and thus they are in a narrow energy gap between the p- and n-layers. The generated electron-holes pairs per quantum well are not many, but because of the small area they are in the probability of recombination is high and thus photons are produced. That is, a small current produces a sufficient amount of coherent photons all within a narrow linewidth. In addition, the very thin quantum well layer causes the released photons to emit perpendicularly to the surface of the layer, and thus these are called *surface emitting lasers* (SEL). In addition, the vertical structure of SELs guides or constrains the photons to an almost cylindrical beam (TEM_{00}) that emerges perpendicular from the top surface of the device, and thus it is called *vertical-cavity surface-emitting laser* (VCSEL), Figure 2.6; the produced almost cylindrical beam has a small divergence (<3 mrad).

The active region sandwiched between two Bragg reflectors comprises a vertical Bragg resonant cavity. The Bragg reflectors and the active region determine the wavelength desired. For example, In+Ga+As+P is used for lasers in the wavelength window

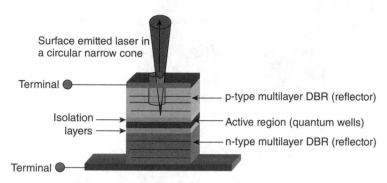

Figure 2.6. A VCSEL emits light perpendicular to its surface.

from 1300 nm to 1550 nm. Other VCSELs are made to emit in the 780 nm to 980 nm range, making them suitable for erbium doped fiber amplifier (EDFA) laser pumps [5] or to emit laser light at 850 nm suitable for multi-mode fiber applications. VCSEL efficiency is relatively satisfactory, 25%, as compared with DFB lasers; for example, a VCSEL may draw 15 mA to produce the same light intensity with a DFB at 60 mA. Because of this, VCSELs do not require temperature control like DFBs do. Direct modulated VCSEL devices at 2.5 Gb/s are common at an output optical power in the range 1 to 1.5 mW; their output depends on device structure, materials, and manufacturer. However, a VCSEL produces lower output power than a DFB and it is not suitable for long haul applications. Because the VCSEL structure is very compact (several micrometers wide), it lends itself to very dense integration and thus cost-effectiveness.

2.2.4.4 Titanium-doped Sapphire Lasers

Crystal-growth advancements have made it possible to create solid state Titanium-doped sapphire (Ti:sapphire) lasers at high-power, better yields and lower cost. The power, yield and cost of solid state lasers depends on the purity of the crystal, and on the size of the artificially grown Ti:sapphire ingot or *boule*. Currently, ingots of 20 cm (200 mm) diameter are reality, which is expected to double in the next few years (a grown ingot has the shape of a large diameter cucumber). Ingots are subsequently sliced to produce large diameter round wafers of about 1–2 mm thin. On a single wafer, many laser devices are manufactured, thus sharing the overall processing cost.

Laser devices are expected to reach petaWatt power level in few femtosecond short pulses. It is also expected that Ti:sapphire lasers will find specific applications in communications (particularly in inter-satellite FSO), in medicine and other fields.

2.2.4.5 Laser Comparison

Different lasers meet different needs and cost-performance models. Lasers may be cooled or uncooled. They may be integrated with modulators or stand-alone. They may be high or low power. They may produce a thin (almost) circular beam or an elliptical one. They may be tunable or fixed wavelength.

They may produce a narrow linewidth or a broad linewidth. They may be suitable for long-haul or for short-haul applications, and so on. Therefore, a fair comparison between different types is not a simple process and it requires expertise.

2.3   MODULATORS

Certain optical materials exhibit some desirable properties: they significantly affect the frequency, or phase, or polarization, or optical intensity of monochromatic light that traverses them, and in a controllable and timely manner. Such materials are used as optical modulators (for different technologies of modulators see [6–8]).

There are two types of modulators, *indirect* and *direct*.

- An indirect modulator is positioned in line with the optical path of a continuous wave (CW) laser, and as the beam traverses the device, one of the light characteristics is affected when applying a modulated electrical signal to the modulator. Modulators may be either *external* to the laser source or *monolithically integrated* with it.
- A direct modulator is the laser device itself when one of the properties of light generated by it is controlled by applying directly a modulated electrical voltage to it. However, because direct modulation at very high bit rates may cause chirp, jitter and noise, the laser device must be carefully designed to minimize such impairments.

The parameters that characterize the performance of optical modulators are *modulation depth, bandwidth, insertion loss, degree of isolation*, and *power*.

Additional parameters are:

- *Modulator sensitivity on polarization*: It refers to modulator's transmissivity and performance based on the polarization state of the modulated optical beam; it is also known as *polarization sensitivity*.
- *Frequency response*: It refers to the frequency at which the output optical power falls at half its maximum value; it is also known as 3 dB modulation bandwidth.
- *Chirp*: It refers to the amount and amplitude of added or enhanced frequency side-lobes to the center frequency of the optical signal.
- *Insertion loss*: It refers to the loss of the optical signal.

2.3.1   The Amplitude Modulation

Analysis of amplitude (intensity) modulation of an optical channel explain certain degradations when very fast bit rates (>10 Gb/s) are used. Here, we present a simple modulation case to identify such degradations.

Consider a monochromatic signal, ω_c, which is amplitude modulated by a function $g(t)$, and described by:

$$m(t) = g(t)\cos\omega_c t \qquad\qquad 2.9$$

The amplitude modulating function $g(t)$ is described in general by:

$$g(t) = [g_0 + mv(t)] \qquad\qquad 2.10$$

and thus,

$$m(t) = [g_0 + mv(t)]\cos\omega_c t \qquad\qquad 2.11$$

where m is the modulation index (equal to 1 for 100% modulation), g_0 is a DC component that for simplicity can be set to 1 and $v(t)$ is the modulating function.

Now, depending on the function of $v(t)$, the modulated signal $m(t)$ may be expanded in three terms:

$$m(t) = g_0\cos\omega_c t + (m/2)[g_m\cos(\omega_c - \omega_m)t] + (m/2)[g_m\cos(\omega_c + \omega_m)t] \qquad 2.12$$

Here we have assumed a simple case where $v(t) = g_m\cos\omega_m t$. The above expansion yields three terms, the main frequency (first term) and two sidebands, each ω_m far from ω_c.

In vectorial representation, the latter relationship is written as:

$$m(t) = \mathrm{Re}[e^{j\omega_c t} + (m/2)e^{j(\omega_c - \omega_m)t} + (m/2)e^{j(\omega_c - \omega_m)t}] \qquad\qquad 2.13$$

where Re denotes the real part of the complex exponential notation. If the e^{xt} term is considered to be a phasor, then $m(t)$ consists of three terms, a *stationary* and two *counter-rotating*, the sum of which yields the modulation signal. If the amplitude of the carrier is unity, then each sideband has a power of $m^2/4$ and both $m^2/2$.

Based on this and under certain worst case conditions an interesting degradation may occur; if the lower sideband frequency ($\omega_c - \omega_m$) is shifted clockwise by θ degrees and the upper sideband ($\omega_c + \omega_m$) clockwise by $180-\theta$ degrees, then, the resultant vector represents a phase-modulated wave, the amplitude modulation of which is largely cancelled, or the modulation index becomes zero. Clearly, in cases in-between, the degradation of the modulation index is partial.

In optical transmission, the two sidebands represent different wavelengths, one at λ_1 corresponding to ($\omega_c - \omega_m$) and another at λ_2 corresponding to ($\omega_c + \omega_m$) which may travel at different speeds if there is dispersion, and thus at different phases. Consequently, On-Off keying (OOK) modulation, under certain conditions, is expected to trigger certain interesting phenomena.

The probability of bit error for Amplitude Shift Keying (ASK) and for the OOK modulation method is given by a formula in terms of the signal level, S, and the noise level, N:

$$\text{ASK (Coherent): } P_e = (1/2)\,\mathrm{erfc}\,\sqrt{[S/(4N)]} \qquad\qquad 2.14$$

TABLE 2.1. BER values correcponding to SNR (dB)

BER	SNR (dB)
10^{-10}	19.4
10^{-9}	18.6
10^{-8}	18
10^{-7}	17.3
10^{-6}	16.4
10^{-5}	15.3

and

$$\text{OOK: } P_e = (1/2) \, \text{erfc} \, \sqrt{(S/N)} \qquad 2.15$$

where erfc is the complimentary error function, the value of which is obtained from mathematical tables. Table 2.1 lists some BER and corresponding SNR approximate values:

Example: Calculate the probability error in the ASK (coherent) case if S/N = 18 dB.
Answer: The S/N power ratio is calculated from 18(dB) = 10 log x as x = 63.36.
Then, the probability error is calculated:

$$P_e = (1/2) \, \text{erfc} \, (\sqrt{[63.36/4]}) = (1/2) \, \text{erfc} \, (\sqrt{(15.84)})$$
$$= (1/2) \, \text{erfc} \, (3.98) = (1/2)[1.8 \times 10^{-8}] = 9 \times 10^{-9}. \qquad 2.16$$

2.4 PHOTODETECTORS AND RECEIVERS

Photodetectors are transducers that alter one of their parameters according to the amount of photons impinging on them. Thus, light may affect conductivity (photoresistors), electrochemical properties (rods and cones of retina), or the amount of generated electron-hole pairs (photodiodes).

Photodiodes are semiconductor devices similar to diodes, but their junction is allowed to be exposed to external light. In semiconductor photodiodes, when a photon interacts with the junction, an electron is excited from the valence to the conduction band, leaving a positive hole in the valence band. However, the photodiode response time and the sensitivity, or the ability to generate hole-electron pairs per photon, differs by type of photodiode.

Semiconductor photodiodes have many advantages and have been deployed in a number of applications. They are cost-efficient, generate measurable current for optical power from picowats to milliwatts, respond to wavelengths from 190 to 2000 nm, respond quickly (as fast as 10 ps), are inexpensive, and can be produced with a small or a large form factor (as large as 10 cm^2). However, at very low optical power levels, thermal noise may be an issue and the amount of signal to noise ratio (SNR) should be considered.

In communications, fast response, spectral responsivity, power sensitivity and low dark current are very important attributes. The two most commonly used materials for photodetectors are Si and indium gallium arsenide (InGaAs) because they have a broad spectral response in the near-IR range, which is suitable to optical communications. Si photodiodes respond in the range of 190 to 1100 nm, whereas InGaAs respond in the range 800 to 1800 nm. Other compounds can respond in the IR range.

Germanium is also suitable and responds to 1600 nm but is not widely used because of high dark current value, which is manifested as noise. Now, because these materials have a broad spectral response and do not select a specific wavelength, in wavelength division multiplexing (WDM) applications [9], externally tuned pass-band filters are used to select specific wavelengths.

In particular, photodetectors with high spectral sensitivity, very fast response time (fast rise and fall time) and with a response in a wavelength range that matches the range of transmitted wavelengths are desirable. Such photodetectors are the semiconductor *positive intrinsic negative* (PIN) photodiode and the *avalanche photo diode* (APD).

The principles of solid state photodetectors are based on the p-n junction potential distribution and the photon energy that penetrates it. For example, when a p-type and an n-type semiconductor material come in electrical contact, then the Fermi level of the two line up and as a consequence the conduction and valence levels of the two are not lined-up at the same level (the p is higher) and a potential difference (or drift space) between the two is created, known as the *depletion layer*. The Fermi level of a p-type semiconductor material is between the conduction and valence layers but closer to the latter; for an n-type, it is closer to the conduction layer. The energy difference between valence and conduction energy bands is known as the *band gap*, and for different crystals, the energy gap has different value; for example, Ge has 0.67 eV, Si 1.12 eV, InP 1.35, and GaAs 1.42 eV. Thus, when an electron is excited in the p-conduction level, it drifts to the n-conduction level as a result of the potential difference between the two. The two levels may be aligned by an externally applied forward bias (voltage), or be further separated by a reverse bias, that is, an external voltage can control the separation and thus conduction.

2.4.1 Cut-off Wavelength

In certain p-n junctions, electrons are excited by photons from the valence to the conduction level. Clearly, the excitation energy must be equal to the energy gap between the valence and conduction bands, Eg. This gap defines the minimum photon energy, or the longest wavelength (known as *critical* or *cut-off wavelength*), above which neither excitation nor absorption takes place, and the junction appears transparent to the photon.

The longest wavelength (or critical wavelength) is derived from the relationship $Eg = h\nu = hc/\lambda$, where $hc = 1.24$ eV.µm and λ is the critical wavelength. Since Eg is not the same for all materials (Ge, In, P, As, Ga, etc.), the critical wavelength depends on the type of materials that comprise the p-n junction. There are several compound materials that can absorb light and thus be used as photodetecting materials, such as ZnSe, GaAs, CdS, InP, InAs, etc. However, in fiber communications, only those materials that absorb photons in the wavelength range 0.8–1.65 µm are usable. As an example, for GaAs, Eg is approximately 1.42 eV, and thus the cut-off wavelength is 1.24 (eVµm)/1.42 (eV) = 0.87 µm. That is, GaAs is transparent to wavelengths greater than 0.87 µm, and thus it cannot detect them; GaAs detects FSO wavelengths shorter than 870 mm, and InGaAs detects longer wavelengths.

As electrons are excited by photons to the conduction level, they drift in the depletion layer. However, these electrons travel at a speed that depends on the potential difference across the drift space (or depletion layer), and on the physical length of the junction, which depends on the lattice constant determined by the atomic arrangement of the compound; the flow of these electrons constitute the photocurrent.

2.4.2 Photodetector Parameters

Based on the above, the most important photodetector parameters for communications are: *spectral response, photosensitivity, quantum efficiency, dark current, forward biased noise, noise equivalent power, terminal capacitance, timing response* (rise time and fall time), *frequency bandwidth*, and *cut-off frequency*. These are defined as follows:

- *Spectral response* relates the amount of current produced with the wavelength of impinging light. Different materials respond differently to electromagnetic radiation. Si detectors respond well to short wavelengths (0.5–1.2 µm), they are compatible with integrated Si devices and they are inexpensive, but they fall short in the DWDM wavelength range 1.3–1.6 µm; whereas InSb responds to the wider spectrum Si 0.5–5.1 µm.
- *Photosensitivity* is the ratio of optical power (in Watts) incident on the device to the resulting current (in Amperes), also known as *responsivity* (measured in A/W).
- *Absolute spectral power responsivity* is the ratio of the output photocurrent of the photodetector (in amperes) to the spectral radiant flux (in watts) at the input of the photodetector.
- *Sensitivity* (in dBm) is the minimum input optical power detected by the receiver (at a certain bit-error-rate (BER)).
- *Quantum efficiency* is the number of generated electron-hole pairs (i.e., current) divided by the number of photons.
- *3-dB bandwidth* is the highest bit-rate at which the output photocurrent swing falls at half its maximum.
- *Dark current* is the amount of current that flows through the photodiode at the absence of any light (hence, dark), when the diode is reverse biased. This is a source of noise under reversed bias conditions.
- *Forward biased noise* is a (current) source of noise that is related to the shunt resistance of the device. The shunt resistance is defined as the ratio voltage (near

0 V) over the amount of current generated. This is also called *shunt resistance* noise.

- *Noise equivalent power* is defined as the amount of light (at a given wavelength) that is equivalent to the noise level of the device.
- *Timing response* of the photodetector is defined as the time lapsed for the output signal to reach from 10% to 90% its amplitude (also known as rise time) and from 90% to 10% (also known as fall time).
- *Terminal capacitance* is the capacitance from the p-n junction of the diode to the connectors of the device; it limits the response time of the photodetector.
- *Frequency bandwidth* is defined as the frequency (or wavelength) range in which the photodetector is sensitive. The frequency sensitivity boundaries are found from the wavelength with maximum power level and at a power drop measured in decibels, such as 3 db, or measured in percentage, such as 10%.
- *Cut-off frequency* is the highest frequency (shortest wavelength) the photodetector is (meaningfully) sensitive.

Based on these definitions, the shot noise current I_{S-N} from a photodetector is expressed by the relationship:

$$I_{S-N} = \sqrt{[2e\,(I_{dark} + I_{ph})B]} \qquad\qquad 2.17$$

where, I_{dark} is the current that flows at the absence of the photonic signal, hence *dark current*, I_{ph} is the photocurrent generated by the photonic signal, and B is the bandwidth of the photodetector (in MHz).

Similarly, when the generated current flows through a load resistsor, thermal noise current is generated, I_{Th-N}, the mean value of which is expressed by:

$$I_{Th-N} = \sqrt{[4kTB/R]} \qquad\qquad 2.18$$

where, T is the temperature (in degrees Kelvin), B is the bandwidth of the photodetector, k is Boltzmann's constant, and R is the load resistance.

If the signal power is expressed in terms of the photocurrent and the load resistor, then the signal to noise ratio, SNR, is expressed by:

$$SNR = 10\log\,[\text{Signal Power/Total noise power}]$$
$$= 10\log\,[I_{Th-N}{}^2/(I_{S-N}{}^2 + I_{ThN}{}^2)]\,(dB) \qquad 2.19$$

This relationship assumes an incoming signal free from noise. In reality, the incoming photonic signal already contains optical noise. As a result, optical noise needs to be included in the total noise power to calculate a realistic SNR.

2.4.3 The PIN Photodiode

The PIN semiconductor photodiode consists of an intrinsic (lightly doped) region sandwiched between a p-type and an n-type. When it is reversed biased, its internal impedance is almost infinite (as an open circuit) and its output current is proportional to the input optical power.

The input-output relationships that define the *responsivity*, R, and the *quantum efficiency*, η, of the photodiode are:

$$R = (\text{output current I})/(\text{input optical power P}) \ (\text{Amperes/Watts}) \qquad 2.20$$

and,

$$η = (\text{Number of output electrons})/(\text{Number of input photons}) \qquad 2.21$$

The quantities R and η are related through the relationship:

$$R = (eη)/(hv) \qquad 2.22$$

where, e is the electron charge, h is Plank's constant, and v is frequency.

When a photon creates an electron-hole pair, the PIN produces a current pulse with duration and shape that depends on the R-C time-constant of the PIN device. The capacitance of the reversed biased PIN photodiode is a limiting factor to its response (and to switching speed).

At low bit rates (<Gb/s), the parasitic inductance of the PIN may be neglected. However, as the bit rate exceeds Gb/s, parasitic inductance becomes significant and causes "shot noise".

2.4.4   The APD Photodiode

The avalanche photodiode (APD) is a semiconductor device, which in operation is equivalent to a photomultiplier. It consists of a two-layer semiconductor sandwich where the upper layer is n-doped and the lower heavily p-doped. At the junction, charge migration (electrons from the n-doped and holes from the p-doped) creates a depletion region. From the distribution of positive and negative charges, a field is created in the direction of the p-layer.

When reverse bias is applied and no light impinges on the device, then due to thermal generation of electrons a current is produced, known as "dark current," which is manifested as noise. If the reversed bias device is exposed to light, then photons reach the p-layer and cause electron-hole pairs. However, because of the strong field in the APD junction, the pair flows through the junction in an accelerated mode. In fact, electrons gain enough energy to cause secondary electron-hole pairs, which in turn cause more, and thus, an *avalanche* process takes place, similar to a photomultiplier, that generates a substantial current. However, the generated electrons build up a charge and if the bias voltage is below the breakdown point, the built-up charge creates a potential that counteracts the avalanche mechanism and thus the avalanche ceases. If the bias is above the breakdown voltage, then the avalanche process continues and a large current is obtained from a single photon.

APDs are made with Si, InGaAs or Ge, and APD structures come in different types:

- The *deep-diffusion* type has a deeper n-layer than the p-layer, and a resistivity so that the breakdown voltage is high, at about 2 KV. Thus, a wide depletion layer is created, more electrons than holes, and a reduced dark current. In general, this type has high gain at wavelengths shorter than 900 nm and switching speed no faster than 10 ns. At longer wavelength (>900 nm), the speed increases but the gain decreases.

- The *reach-through* type has a narrow junction and thus photons travel a very short distance until they are absorbed by the p-type to generate electron-hole pairs. These devices have uniform gain, low noise and fast response.

- The *super-ionization* type is similar to reach-through but with a structure so that the accelerating field is gradually increasing. As a result, a low ionization number of holes to electrons ratio (for certain amount of incident light) is achieved that increases the gain, the carrier mobility (electrons are faster than holes), and the switching speed.

During this multiplication (avalanche) process, the shot-noise is also multiplied and it is estimated to:

$$\text{Shot-noise} = 2eIG^2F \qquad\qquad 2.23$$

where F is the APD noise factor and G is the APD gain expressed as:

$$G = I_{APD}/I_{primary} \qquad\qquad 2.24$$

I_{APD} is the APD output current and $I_{primary}$ is the current due to photon-electron conversion.

If τ is the effective transit time through the avalanche region, the APD bandwidth is then approximated to:

$$B_{APD} = 1/(2\pi G\tau) \qquad\qquad 2.25$$

The APD output signal current, I_o, is

$$I_o = G * R_o * P_i, \qquad\qquad 2.26$$

where R_O is the intrinsic responsivity of the APD at a gain G = 1 and wavelength λ, G is the APD gain, and P_i is the incident optical power. The gain is a function of the APDs reverse voltage, V_R, and it varies with applied bias.

APDs may be made with silicon, with germanium, or with indium-gallium-arsenide. However, all types do not have the same performance and characteristics. The type of material determines the responsivity, gain, noise and switching speed of the device. For example, indium-gallium-arsenide responds well in the range 900 to 1700 nm (suitable in communications), and has low noise, fast switching speeds, but is relatively expensive. Silicon responds in the range 400 to 1100 nm, is very inexpensive and is easily integrated with other silicon devices; germanium is a compromise between these two.

2.4.5 Photodetector Figure of Merit

The following is a summary of three figures of merit (FOM) that are important in the performance evaluation of photodetectors. Additional specific figures of merit may also be provided by photodetector manufacturers.

- Responsivity (R) = V_s/(H A_d) (VW^{-1})
- Noise Equivalent Power (NEP) = HA_d {V_n/V_s) (W)
- Detectivity (D) = 1/(NEP) (W^{-1})

Where V_s is the root mean square (rms) signal voltage (V), V_n is the rms noise voltage (V), A_d is the detector area (cm^2), and H is the irradiance (Wcm^{-2}).

In general, APD photodetectors have a higher gain than PIN photodetectors, but PINs have a much faster switching speed and thus they have been largely deployed in high bit rate detection (40 Gb/s), particularly the waveguide-type PIN detectors. However, APDs are improving to combine their high gain advantage with fast switching speeds.

2.4.6 Silicon and InGaAS Photodetectors

2.4.6.1 Silicon-based Photodetectors Si is used in photodetectors in the visible and near-IR wavelength range, at low levels of light, and at high switching speeds suitable in high data rates (10 Gb/s). Because many other optical components are made of Si, Si-based photodetectors are popular because of component integrability for low cost applications, such as 10 GbE.

Si-based detectors have a typical spectral response in the 300 to 1100 nm range with maximum sensitivity around 850 nm, and thus they are suitable with short-wavelength VCSELs. However, the Si sensitivity at wavelengths above 1000 nm drops off dramatically, with a cut-off wavelength at 1100 nm.

For lower-bandwidth applications (1 Gb/s), Silicon PIN (Si-PIN) and Silicon APD (Si-APD) detectors are popular. There are also Si-PIN detectors with integrated transimpedance amplifiers (TIAs).

Si-PIN detector sensitivity is a function of signal modulation bandwidth, which decreases as the detection bandwidth increases. A typical sensitivity value for Si-PIN photodiodes is around −34 dBm at 155 Mb/s.

Si-APDs are far more sensitive, due to internal amplification (avalanche). Therefore, Si-APD detectors are better suited to FSO systems. A Si-APD sensitivity value for higher-bandwidth applications is around −52 dBm at 155 Mb/s.

2.4.6.2 InGaAs-based Photodetectors InGaAs is also a wideband detector material that is used in communications. Contrary to Si, InGaAs detectors have a superior response in the range from below 1300 to above 1620 nm with maximum at 1550 nm, as well as low noise current; as such, nearly most fiber-optic systems use InGaAs as a detector material.

InGaAs receivers are based on PIN or APD. Between the two, InGaAs APDs are far more sensitive due to internal amplification (avalanche). Typical sensitivity value for InGaAs APD is around −46 dBm at 155 Mb/s, and around −36 dBm at 1.25 Gb/s.

Very high data rates (10 and 40 Gb/s) are also possible with smaller footprint InGaAs APD detectors that have a shorter junction and shorter carrier travel time. However, as the device size becomes smaller, so does the device aperture, which makes light coupling more challenging.

2.4.6.3 Ge-based Photodetectors

Ge-APDs have a response from 800 to 1600 nm, but they have a higher noise current than InGaAs-APDs. However, Ge-APDs are very attractive to applications that demand higher speed and higher sensitivity than PINs.

2.4.6.4 Selecting a Photodetector

Based on the aforementioned, the choices for the best photodetector are many; Silicon, Germanium or InGaAs, PIN ro APD, offers operation at a variety of wavelengths used in Free Space Optics. Which photodetector is the best is predicated by specific application requirements.

In general, APDs offer higher sensitivity than PINs at medium-to-high bandwidths.

In communications, sensitivity and low spectral noise are important, as is spectral response and gain. A gain from 100 to 1000 for Si-APDs, or 10 to 40 for Ge-APDs and for InGaAs-APDs is achievable.

Comparing APDs with PINs at the same quantum efficiency, PIN detectors have better signal-to-noise ratio (SNR) than APDs.

In short, the selection of the pair transmitter-detector plays a critical role in the design of the FSO transceiver and FSO link performance.

2.5   OPTICAL AMPLIFICATION

Point-to-point FSO links of a relatively short length (<2 Km) require no optical amplification on the link since the link has been budgeted to meet the photodetector sensitivity for the laser emitted power, considering the medium attenuation under most usual conditions and other losses (connectors, etc. However, in very long links, one or more FSO relays may be required to amplify the optical signal until it reaches its end destination. In such case, amplification may be accomplished by using an optical-to electrical to optical (O-E-O) amplification system or an optical to optical (O-O) amplification system.

O-E-O amplification is a multistep process. Traditionally, a O-E-O first converts the optical signal to electronic, the electronic signal is retimed, reshaped and amplified (an operation known as 3R), and then it is converted back to optical. This is also known as *regeneration*. However, although the O-E-O regenerator is technically more straightforward, as compared with the O-O amplifier, it is costlier.

Alternatively, direct *optical amplifiers* (OA), directly amplify a weak optical signal but they do not perform all 3R functions. To date, the two best known amplification methods applicable to FSO are *semiconductor optical amplifiers* (SOA) and *optical fiber amplifiers* (OFA).

2.5.1 Optical Amplifier Characteristics

The key common characteristics of optical amplifiers are *gain, gain efficiency, gain bandwidth, gain saturation* and *noise, polarization sensitivity* and *output saturation power*. Other characteristics are *sensitivity (gain and spectral response) to temperature* and other environmental conditions, *dynamic range, cross-talk, noise figure, physical size* and others.

- *Gain* is the ratio of output power to input power (measured in dB).
- *Gain efficiency* is the gain as a function of input power (dB/mW).
- *Bandwidth* is a function of frequency, and as such *gain bandwidth* is the range of frequencies over which the amplifier is effective.
- *Gain saturation* is the maximum output power of the amplifier, beyond which it cannot increase despite the input power increase.
- *Noise* is an inherent characteristic of amplifiers. In electronic amplifiers, noise is due to (random) spontaneous recombination of electron-hole pairs that produces an undesired signal added to the information signal to be amplified. In optical amplifiers, it is due to spontaneous light emission of excited ions, which we will further explore.
- *Polarization sensitivity* is the gain dependence of optical amplifiers on the polarization of the signal.
- *Output saturation power* is defined as the output power level for which the amplifier gain has dropped by 3 dB.

Based on principle of operation, OAs are distinguished in:

- Semiconductor optical amplifiers (SOA)
- Optical fiber amplifiers (OFA)
- Stimulated Raman amplifiers (not applicable to FSO)
- Stimulated Brillouin amplifiers (not applicable to FSO)

In this section we examine the O-O amplifiers, the semiconductor optical amplifier and the doped fiber amplifier [10–12].

2.5.2 Semiconductor Optical Amplifiers

Semiconductor optical amplifiers (SOA) are based on conventional laser principles; an active waveguide region is sandwiched between a p- and a n-region. A bias voltage is applied to excite ions in the region and to create electron-hole pairs. Then, as light of a specific wavelength is coupled in the active waveguide, stimulation takes place and causes electron-hole pairs to recombine and generate more photons (of the same wavelength with the optical signal), and hence optical amplification is achieved. For best coupling efficiency of the optical signal in the active region, the end walls of the SOA have been coated with an anti-reflecting material.

The excitation and recombination of electron-hole process is described by rate equations. However, the rate of electron-hole generation and the rate of recombination must be balanced for sustained amplification. This depends on many parameters, largely by the active region and the bias voltage, as well as the density and the lifetime of carriers. In the presence of a photonic signal, the number of recombined electron-hole pairs per stimulating photon provides a direct measure of the spectral response and of the optical power and thus the spectral gain of the SOA. The amplifier gain, G, is approximated to

$$G = -12[(\lambda - \lambda_p)/\Delta\lambda]^2 + G_p \qquad\qquad 2.27$$

where G_p is the peak gain at the corresponding wavelength λ_p, $\Delta\lambda$ is the full-width half-maximum (FWHM) gain bandwidth, and the factor 12 is a result of the definition of $\Delta\lambda$.

The 3-db saturation output power, P_s, is described as a function of λ by:

$$P_s = q_s(\lambda - \lambda_p) + P_{s-p} \qquad\qquad 2.28$$

where q_s is a linear coefficient and P_{s-p} is the 3-db saturation output at the peak gain wavelength λ_p.

Depending on structure, SOAs are distinguished in:

- Semiconductor traveling wave laser optical amplifiers
- Fabry-Perot laser amplifiers, and
- Injection current distributed feedback (DFB) laser amplifiers

The SOA salient characteristics are:

- High gain (25–30 dB)
- Output saturation power in the range of 5 to +13 dBm
- Nonlinear distortions
- Wide bandwidth
- Spectral response in the wavelength regions 0.8, 1.3, and 1.5 μm
- SOAs are made with InGaAsP and thus they are small, compact semiconductors easily integrable with other semiconductor and optical components.
- SOAs may be integrated into arrays
- Polarization dependency; thus, they require a polarization-maintaining fiber (polarization sensitivity 0.5 to1 dB)
- Higher noise figure than EDFAs (higher than 6 dB over 50 nm)

- Higher cross-talk level than EDFAs due to non-linear phenomena (four-wave mixing).

Because SOAs are compact solid state devices, their fast responding non-linearity may also be employed in wavelength conversion, regeneration, and other functions.

2.5.3 Optical Fiber Amplifiers

Optical fiber amplifiers (OFA) are specialty fibers that have been heavily doped with one or more rare-earth elements. Their purpose is to absorb optical energy from one spectral range and emit optical energy (or fluoresce) in another, and especially in the range useful in fiber communications (800–900 or 1300–1620 nm). However, each element has its own absorption-emission characteristics. For example, an OFA does not amplify any optical signal but signals within a specific range that is determined by the dopant type, and also does not exhibit exactly the same gain for all wavelengths in the amplification spectral range; the latter defines the degree of *gain flatness* of an OFA.

In general, for sustained amplification, the rate of excitation should be less or equal to the rate of stimulation plus the rate of spontaneous emission:

$$dN_e/dt \leq dN_{st}/dt + dN_{sp}/dt \qquad 2.29$$

where N_e is the number of excited electrons to a higher energy, N_{st} the number of electrons returning to lower energy by stimulation, and N_{sp}, spontaneously. That is, the excitation/stimulation and amplification process reduces to a flow problem described by a set of differential rate equations.

Some of the attractive rare-earth elements that are used in optical fiber amplification are Nd^{3+} and Er^{3+} that emit in the ranges 1.3 and 1.5 μm, respectively. Besides Er and Nd, other rare-earth elements have been used such as Ho, Te, Th, Tm, Yb, and Pr, or in combination (e.g. Er/Yb), each operating in various spectral bands, Figure 2.7.

2.5.4 Erbium Doped Fiber Amplifiers

The *erbium doped fiber amplifier* (EDFA) is one of the most popular in optical communications because its spectral emission, 1530 to 1565 nm, matches the C-band in

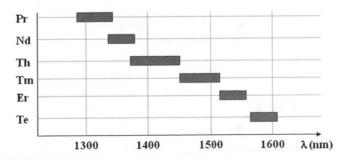

Figure 2.7. Different elements may be used in DFAs to amplify corresponding optical bands.

DWDM, and because erbium is excited by several optical frequencies, known as *pumps*, at a wavelength of 514 nm, 532 nm, 667 nm, 800 nm, 980 nm and 1480 nm; the last two are the most popular pumps in EDFAs used in fiber-optic communications.

The shortest wavelength, 514 nm, excites erbium to the highest possible energy level, from which it drops to one of four intermediate metastable levels radiating phonons (the acoustical quantum equivalent of photon). A similar quantum activity takes place with the remaining excitation wavelengths, and the longest wavelength, 1480 nm, excites erbium to the lowest metastable energy level. When at the lowest metastable level, the excited erbium drops to the initial (ground) level emitting photons of a wavelength in the range 1530 to 1565 nm.

What is important is that when a photon with wavelength in the range 1530 to 1565 nm passes through the excited EDFA, it stimulates the excited Erbium atoms, which from the lowest metastable level emit photons of the same wavelength with the passing through photon. The result is more photons out than photons in, and thus *light amplification by stimulated emission* occurs. What is also more important is that when two photons with different wavelengths (but in the C-band range) pass through the excited EDFA, both photons stimulate excited erbium atoms and both are amplified. However, when many wavelengths are amplified by the same EDFA, the total gain of the EDFA is shared among the stimulating photon-wavelengths. In such a case, however, if the stimulation process is faster than the excitation process, then the EDFA may be quickly depleted, in which case the amplification process becomes unstable.

Now, when erbium ions are excited by a 980 nm pump, after approximately 1 μs the excited ions fall on the lowest metastable energy level from which, if triggered (or stimulated), they drop to ground energy level emitting photons that matches the wavelength of the triggering photon. However, if they are not triggered, then, after approximately 10 ms (known as *spontaneous lifetime*), they spontaneously drop from the lowest metastable level to the ground level emitting photons in the range around 1550 nm, Figure 2.8. Clearly, in optical communications such spontaneous emission could be manifested as additive optical noise. Conversely, because the bit period at 1 Gb/s is very short (ns) as compared with the spontaneous lifetime of erbium (ms), and because stimulation of excited erbium atoms is almost instantaneous, inter-symbol interference is not an issue.

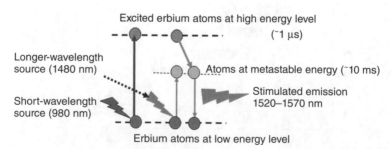

Figure 2.8. Excitation levels for Erbium.

Some of the important parameters in the EDFA are:

- Concentration of dopants
- Relative population of upper states
- Lifetime at upper states
- Effective area of EDFA fiber
- Length of EDFA fiber
- Absorption coefficient
- Emission coefficient
- Power of pump
- Power of signal
- Number of wavelengths (channels) in the signal
- Direction of signal propagation with respect to pump flow

EDFAs with gain greater than 50 dB have already been applied in fiber-optic communications, as well as in FSO relay stations that operate in the C-band. However, because FSO signals do not carry many optical channels (each at different wavelength) but few, the EDFA gain is not shared by many channels as in fiber-optic communications and therefore there is no issue with depleting the EDFA. Therefore, a low gain and low noise characteristics EDFA may be used in FSO relay stations, increasing signal quality and decreasing cost.

EDFA advantages:
- EDFAs have a high power transfer efficiency from pump to signal (>50%)
- They directly and simultaneously amplify a wide wavelength region over 80 nm (in the region of 1550 nm), at an output power as high as +37 dBm, with a relatively flat gain (>20 dB), which is suitable to WDM systems. Modified EDFAs can also operate in the L-band.
- The saturation output is greater than 1 mW (10–25 dBm)
- The gain time constant is long (>100 ms) to overcome patterning effects and inter-modulation distortions (low noise)
- They are transparent to optical modulation format
- They have a large dynamic range
- They have a low noise figure
- They are polarization independent (thus, reducing coupling loss to transmission fiber), and,
- They are suitable for long-haul applications

EDFA disadvantages:
- They are not small devices (they are kilometer-long fibers) and they cannot be integrated with other semiconductor devices.

- EDFAs exhibit amplified spontaneous light emission (ASE). That is, even if no incoming optical signal is present, there is always some output signal as a result of some excited ions in the fiber; this output is termed spontaneous noise.
- Other drawbacks are crosstalk, and,
- Gain saturation.

Erbium has also been (experimentally) used to dope solid state waveguides to produce compact optical amplifier devices, called Erbium doped waveguide amplifier (EDWA), which may also be integrated with couplers and combiners on the same substrate to yield small form-factor components.

2.6 OPTICAL SIGNAL TO NOISE RATIO

2.6.1 Signal Quality

If a signal would suffer by attenuation only and everything else was perfect, then its diminished power would be easy to deal with. Unfortunately, many other parameters cause additional signal variations, spectrally and temporally, that are manifested as signal disturbances that deteriorate the quality and integrity of the optical signal.

We have talked about optical noise that may contaminate the optical signal. What is more important is the amount of noise power in the signal power, and specifically, the signal to noise ratio and (in optical communication) the *optical signal to noise ratio* (ONSR). SNR and OSNR is an important signal performance parameter, which is related to bit error rate (BER).

For single mode fiber transmission, given an OSNR in dB, a first order approximation empirical formula to calculate BER is:

$$\log_{10} BER = 1.7 - 1.45 \, (OSNR) \qquad\qquad 2.30$$

Example: Assume that OSNR = 14.5 dB. Then logBER = 10.3, and BER = 10^{-10}

When engineering a FSO link, the optical power that impinges on the aperture of the photodetector needs to be equal or higher than its sensitivity; the optical power at the receiver depends on several parameters, such as:

- Laser device characteristics
- Laser beam characteristics
- Data bit rate
- Wavelength of channel
- Modulation method
- Link length
- Target or expected BER
- Transmitter and receiver aging margin
- Receiver decision threshold margin
- Receiver cross-talk margin

- Receiver gain and noise
- Optical component attenuation or gain (due to beam focusing)
- Transmission medium parameters and effects
- Dynamic and min-max parameter variation of medium
- Optical reflections
- Amplification noise

In communications, four major contributors affect the signal quality: *intersymbol interference* (ISI), *cross-talk*, *noise* and *bit error rate* (BER).

Bit spreading and sidetones in the propagating optical signal may drift in the next symbol and thus two contiguous symbols in the original bit, say "10", they may appear at the receiver as "11" or as "00". This is known as *intersymbol interference* (ISI) and pertains only to an optical channel.

Bits from one optical channel may influence bits in another channel, because of possible interactions or of spectral overlap, so that two contiguous symbols in the original bit, say "10", they may appear at the receiver as "11" or as "00". This is *cross-talk* and it pertains to more than one optical channels in a fiber.

In FSO communication laser sources, optical amplifiers and the unstable medium contribute optical noise that affect OSNR and thus BER.

2.6.2   Signal Quality Monitoring Methods

In order to monitor the signal quality at the receiver, the *termination*, the *sampling*, the *spectral monitoring*, or the *indirect* method may be used.

- The termination method consists of error detecting codes (EDC) that are incorporated in the bit stream at the source. At the receiver, the EDC code calculates the number of erroneous bits in the stream and is capable to correct some.
- The sampling method requires discrete sampling and a signal analyzer that consists at minimum by a demultiplexer, very low noise detectors, a synchronizer, and algorithmic discrete signal analysis.
- The spectral monitoring method consists of noise level measurements and spectral analysis.
- The indirect signal monitoring method depends on information deduced from system alarms (loss of frame, loss of synchronization, loss of signal power, and so on).

2.7   ACQUISITION, POINTING AND TRACKING

FSO transceivers located at either end of a link need to be aligned. Alignment may be achieved manually or semi-manually. Manual alignment involves a telescope so that an operator finds and places the target transceiver at the other end of the link on the cross-hair. A similar process takes place at the other end of the link, and thus the two transceivers are eventually aligned.

Initial alignment is relatively easy with two operators and may be achieved within minutes. However, after the initial alignment, wind, natural phenomena, and building thermal expansion/contraction may displace one or both transceivers from their initial position and affect alignment. If displacement is a few degrees, then the laser from one or both transceivers mis-directs its beam and alignment may be lost. For example, 1 mrad angle of divergence over 1 Km link length produces a beam diameter of about 1 meter ($1° ≈ 17$ mrad and 1 mrad $≈ 0.0573°$). For small angles, the relation {Beam angle (milliradians) $\times$ Link lenth (Km) = beam diameter (m)} can be used.

To continue alignment, an automatic tracking mechanism with associated protocol and software is required at each transceiver. In general, to locate, align and maintain alignment, takes place in three operational phases: Acquisition, Pointing, and Tracking (APT).

2.7.1   Acquisition

The acquisition phase takes place when the two terminals at either end of a link try to locate each other for the first time. Typically, acquisition can be accomplished with two operators, each positioned at either end-point of a link using a telescope and with knowledge of the coordinates of the nodes or of the spatial direction of the link; straight paths and line of sight in FSO networks is a requirement here.

In some cases however, it is not possible to have two operators but only one or none, as in combat fields, in aircraft to tower communication, and in satellite to satellite communication. In the latter case, an auto-acquisition method is used. Thus, we distinguish the following three acquisition cases:

- Stationary: both end-points of the link are stationary
- Semi-stationary: one end-point of the link is stationary and the other is mobile
- Mobile: both end-points of the link are moving.

In the acquisition phase, the key parameter is *acquisition time*, which greatly depends on application method, and on technology; acquisition time may be in the order of a minute to many minutes but less than an hour.

In stationary systems, the simplest acquisition method requires knowledge of topology data; a coordinate frame of reference (for terrestrial applications, a Cartesian frame would suffice), and a vector that describes the orientation in the frame of reference and the distance of the link. This knowledge is used by one or two operators assisted by the relatively short link length (up to few kilometers) and line-of sight (LoS) in a clear day.

In semi-stationary systems, acquisition is challenging because of the moving end. Acquiring the stationary end by the moving end is simpler because the coordinates of the stationary node are fixed. However, acquiring the moving end by the stationary is more challenging since the latter has to find a "moving target", which may not move on a plane, with constant velocity, or in a prescribed trajectory, as in a combat field with uneven terrain. In this case, more sophisticated methods are used, such as the (differential) Global Positioning Systems (GPS) and the Inertial Navigation System (INS), along with a secure RF microwave link. Now, modern GPS systems have a resolution of about ±1 meter, and the laser beam a geometrical spreading at 1–2 Km distance in the same order of magnitude, which turns out to be helpful in the acquisition phase; laser beams with 1 mrad divergence angle, at 1 Km distance yield a cross-sectional surface area of about 1 meter.

In mobile systems, acquisition is most challenging at both end-points of the link. In this case, the (differential) Global Positioning Systems (GPS) and the Inertial Navigation System (INS), along with a secure RF microwave link are necessary at each end-point. However, because two moving nodes in the three-dimensional space may take a long time to achieve acquisition, a third stationary node that communicates with both mobile stations over RF microwave links may be used to assist the acquisition and the pointing process. The third stationary node tracks the two mobile nodes, locates their coordinates in space, identifies direction of motion or trajectory of each node in space, and it sends this information to both mobile nodes. When the two mobile nodes have accomplished acquisition, pointing and auto-tracking, then the assistance of the third stationary node becomes redundant. Again, although modern GPS systems have a resolution of about a meter (±1 m), the laser divergence or geometrical spreading over many kilometers (in satellite to satellite communication) should be of comparable magnitude, which implies that the laser beam should be very narrow; if the laser diverges too much, the geometrical spreading loss is high, the optical power at the receiver is extremely low and it may be near or lower than the sensitivity of the photodetector; this increasies the probability of bit errors and the bit error rate (BER).

In mobile systems, initially a divergent laser *beacon beam*, also known as *probe beam*, may be used during the acquisition and pointing phases. A possible scenario in this case is that, node A slowly scans an area with its beacon beam and in the approximate direction where node B is supposed to be, and eventually the beacon beam comes within the field of view of node B. When the beacon beam is detected by node B, it either sends back to node A its reflection by a corner cube reflector or it sends a narrower beacon beam; a corner cube reflector reflects back in the same direction of the incident light. Node A then either narrows the divergent beacon beam, or sends back to node B a communication beam, and eventually acquisition is achieved; after this and the pointing phase is completed, the beacon beam is switched-off. A similar acquisition process is used between the telecommunications ARTEMIS geostationary satellite (owned by the European space agency ESA) and the OICETS satellite (of the Japan aerospace exploration agency JAXA). Clearly, mobile satellite systems have also their own challenges, vis-a-vis geostationary satellite networks that seem stationary around the earth and over the equator, and low earth orbit satellite systems (LEOS) that move on inclined orbits around the earth in clusters [13]. In mobile terrestrial systems, the acquisition time is critical particularly when the moving nodes are not on a plane but on uneven terrain and they move in non-deterministic paths.

2.7.2 Pointing

When the acquisition phase is completed, the transmitter has sufficient knowledge of the position of the receiver at the other end-point and the pointing phase begins.

Pointing is the operation of directing the communication beam accurately towards the receiver; this may or may not require some fine adjustments, the accuracy of which is known as pointing resolution. Clearly, as the pointing resolution increases, the power efficiency of the transmitted communications beam increases, considering that the beam profile is more intense at its axis of symmetry. Pointing resolution is measured in degrees of angle, θ, and it should be less than half of the divergence angle of the beam. Typical pointing resolutions are from 1 to 200 μrad (0.00002865 to 0.00573 degrees) [14, 15]. The pointing resolution of the ARTEMIS satellite is stated as 1 arcsecond (~0.000278 degree), half of its beam divergence angle [16, 17].

Depending on case, pointing has to overcome different challenges. For example:

- In stationary systems, pointing may have to deal with possible strong winds that move the position of the target receiver due to building sway; small sway may have no significant effect due to the small divergence of the beam.
- In semi-stationary systems, pointing may have to deal with additional issues associated with the movement of the mobile node on uneven terrain and also with wind.
- In mobile systems, the pointing phase may have to deal with movements due to uneven terrain with not-deterministic paths. In mobile satellite networks, it may be simpler between geostationary satellites, or it may be more difficult in LEOS. Again the speed to accomplish the pointing phase is important. Once pointing is completed and the communication beam is locked-in position, then the self-tracking phase will maintain connectivity between the two transceivers on the link.

2.7.3 Tracking

After acquisition and pointing, the communication beam connects the transmitter of node A with the receiver of node B, and vice versa. However, because one of both nodes may move or shift position due to movement or external forces (such as wind), there should be a mechanism to ensure that the direction of the beam is kept continuously towards the receiver and that alignment is maintained. This mechanism is known as tracking, self-tracking or auto-tracking, and it is needed by stationary, semi-stationary and mobile systems, although the issues associated with tracking differ among them.

Tracking is a closed loop system. For example, node A sends a beam to node B. Now, if node A or node B, or both move their relative position by as little as a few meters, then the receiver at node B "sees" a degraded signal power but node A does not know it. However, if node B communicates to node A the signal power or performance it receives from it periodically, and also node A communicates to node B its signal power or performance with the same period, then as soon as the signal degradation occurs and is communicated, one or both nodes may initiate self-correcting align-ment, which actually is an optimization procedure. Thus, some of the requirements are:

- Tracking frequency: How frequently the signal power or performance is com-municated over the loop?
- Tracking resolution: What is the performance degradation step at which self-tracking should be initiated?
- Tracking speed: How fast the beam deviation has been corrected?
- Tracker axes of freedom: Can the tracking mechanism perform correction on two or three dimensions?
- Tracker range of correction: what is the maximum angle within which the track-ing mechanism can move?
- Tracking mechanism ability to distinguish between change of relative position and fog, snow or rain.

Clearly, the impact of the tracking mechanism on communications varies with system. For example, in slow moving nodes, the tracking frequency may be relatively slow, although the other parameters may be more stringent if the terrain is very uneven. Conversely, in very fast moving nodes, tracking frequency may be very critical as well; this also depends on if the nodes move in the same direction, in opposite directions or in an angle with respect to one another. Current tracking mechanisms are able to main-tain beam alignment under strong winds (30–40 mph).

In general, typical tracking mechanisms have the laser mounted on a gimbal, which is controlled with digital servos. Alternatively, the laser may be fixed and a mirror may be mounted on a gimbal. Moving the gimbal steers the optical beam in the three-dimensional space. However, a gimbal is a mechanical device that may also introduce mechanical jitter as it rotates, which is manifested as beam spatial jitter. Thus, the gimbal is suitable for coarse steering. Modern FSO tracking mechanisms involve no gimbals, but electrooptic or acoustooptic devices that are especially applicable to fast moving nodes, such as airplanes and satellites. At the receiving end, an array of pho-todetectors detects movement as well as direction of movement (left-right); a matrix of photodetectors that acts like a Whitestone bridge detects movement on a plane (left-right, and up-down) [18, 19]. When misalignment occurs, the Whitestone Bridge pro-duces a directional error, positive or negative, and the speed of error correlates with the speed of misalignment; the error is minimized as re-alignment occurs.

Now, because the metric that triggers self-tracking is the degradation of the received signal power, or its attenuation, it is important that a tracking mechanism is able to distinguish between attenuation due to misalignment and attenuation due to fog, snow or rain; current auto-tracking mechanisms are able to do this.

2.7.4 Corner Cube Reflectors

In many optical applications, including FSO systems and networks, the incident beam needs to be reflected back in the same direction; a *retroreflector* accomplishes just that

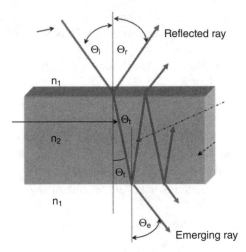

Figure 2.9. Plate reflection and defraction.

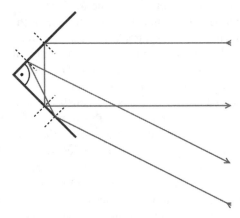

Figure 2.10. The workings of a corner cube retroreflector.

(note: a mirror reflects back in the same direction only if the angle of incidence is zero; that is the optical beam is perpendicular to its flat surface, else it is reflected in an angle away from the angle of incidence, Figure 2.9).

In optical applications, a retroreflector is a passive optical component that may be simply constructed with an optically transparent cube, thus called corner cube reflector, the operation of which is shown in Figure 2.10.

A corner cube reflector has three intersecting surfaces and it works for a relatively wide-angle field of view. Corner cube reflectors may be used in the optical range as well as in the microwave range; they have been used as nautical navigation markers, in radars, in surveying, in vehicle reflectors, in road-signs, in reflecting paint (in which microscopic corner reflector structures are incorporated), in missiles, in satellites, and other applications.

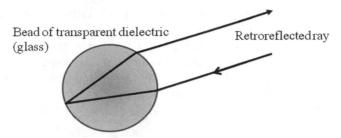

Figure 2.11. Microspheres (beads) construct a retroreflector.

The surface flatness, angle tolerance, and reflecting wavelength are key parameters in optical retroreflectors.

The principle of corner cube reflectors may also be used in acoustic applications. Other techniques have employed micro-electro-mechanical systems (MEMS) [20] as well as microspheres (beads), Figure 2.11, to implement retroreflectors [21].

2.8 ADAPTIVE AND ACTIVE OPTICS

The optical wavefront that propagates in an ideal medium should remain spherical. However, as light propagates through the atmosphere and depending on medium conditions, particularly temperature and turbulence, its ideal wavefront is distorted to a "wrinkled" wavefront that produces blurring and distortion.

In optical communications, distorted wavefronts become a source of noise. In specific applications, this type of noise is a major nuisance that should be overcome, such as astronomy (in astronomical telescopes), satellite to earth FSO communications and retinal imaging systems.

To assess the quality of "seeing" in the presence of atmospheric turbulence and to assess the performance of any adaptive optical correction system, the *Strehl ratio* is used. The Strehl ratio is defined as the ratio of the observed peak intensity at the detection plane from a point source (a laser) compared to the theoretical maximum peak intensity of a perfect imaging system working at the diffraction limit [22, 23].

The technology that can ameliorate distorted optical wavefronts, and thus improve the performance of optical systems, is known as *adaptive optics* (AO) and as *active optics* (AO). Because both have the same acronym, in this book we distinguish between the two as AdO for adaptive optics, and AcO for active optics.

Systems that incorporate adaptive optics or active optics reduce the distortion amount of wavefronts. To accomplish this, they measure the distortion in the wavefront and compensate for it with a *spatial phase modulator*. How fast they compensate for distortion, or the *response time*, distinguishes AdO systems and AcO systems, the former being the fastest. Because of slow response time, active optics are used in slow deformation situations (<0.01 Hz) such as very large telescopes (with a mirror about 4 m diameter), the mirror of which starts slowly changing geometry (its sags) due to gravity as the telescope tilts, as in the Galileo, Gemini and other telescopes.

In general, AdO methods compensate for wavefront distortions by measuring the amount of distortion, actually by measuring the amount of phase distortion, and by making appropriate changes on the surface of a flat mirror; these changes are commensurate with the amount of phase distortion and in a direction that compensates for it. Note that the atmosphere causes phase distortion than amplitude distortion.

A spatial phase modulator can be made with one of several techniques such as, a deformable mirror, with a micro-electromechanical-system (MEMS), or a liquid crystal array.

2.8.1 Methods for Adaptive Optics and Active Optics

The *deformable mirror* (DM) consists of a flexible flat or planar mirror, membrane-like, which by means of a matrix of actuators behind.

Consider a distorted wavefront impinging in an angle on the planar mirror, which reflects the distorted wavefront as is.

Now, if the planar mirror was deformed appropriately, then the distorted wavefront would be smoothed out (wrinkle free), Figure 2.12. In order to deform the mirror, a flexible mirror is used with a matrix of actuators connected behind it, so that, when activated, each actuator "pushes" or "pulls" at the point of contact with the mirror to induce surface deformations that commensurate with the amount of wavefront distortion; the amount and deformation resolution and speed of deformation depends on the

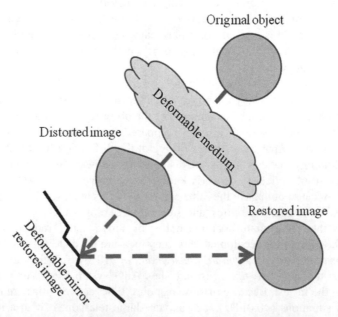

Figure 2.12. A distorted image impinges on appropriately deformed mirror to produce a restored image.

number of actuators, on the actuator pitch and on the elasticity of the mirror. A computer is also used that performs calculations and controls each actuator, at a rate of about 1000 Hz, so that each one induces an appropriate deformation amount.

The wavefront distortion and the corrections to be made are made with a curvature sensor and a waveform analyzer; such sensors may be a simple interferometer or a Shack-Hartmann wavefront sensor. The Shack-Hartmann AdO employs an array of lenslets, whereby the focal point of each lenslet changes by the amount of wavefront distortion that impinges on it [24].

- A **micro-electromechanical system** (MEMS) consist of a large matrix of micro-mirrors (as large as 1000×1000) to form a uniform mirror surface. However, each micro-mirror is individually controlled and it can be electrostatically tilted to reflect rays in different direction. Thus, the mirror is deformed discreetly, it takes place in tiny increments by each micromirror, and thus the overall surface of the MEMS mirror appears to be deformed.
- **Liquid crystal** (LC) is a technology that has been successfully employed in many applications, particularly in flat displays. A LC consists of a nematic liquid, the molecules of which are in random orientation, if no electrical field is applied; in this case, light is blocked from passing through. However, when a field is applied, the long (nematic) molecules are aligned and they allow light passing through, the amount of which, or the phase of which depends on the amount of orientation of the molecules and the type of molecules in the LC. Now, if a LC material is organized in a pixel matrix, so that the molecules in each pixel can be oriented differently, one can control the phase of the distorted wavefront easily [25, 26].

2.8.2 The Reference Star

In order to quickly compensate for distorted wavefronts, due to atmospheric turbulence and temperature variations [27, 28], the priori knowledge of the undistorted wavefront could be useful in deep space communications [29]. That is, if a reference source existed, it could be helpful to measure deviations from it by means of adaptive optics.

In astronomy, a laser probe beam with known profile may be transmitted from a space vehicle to an observatory, which receives the actual distorted profile and compares it with the expected profile to measure distortion. This information is then used to correct images that are observed in space. A different application takes advantage of the fluorescent property of sodium atoms in the mesosphere; in this case, a laser beam from an earth station is directed to the mesosphere where it excites sodium atoms, which fluoresce and thus produce an artificial star. Moreover, a laser probe beam with known profile can be transmitted from an earth observatory to a space observatory where distortion measurements of the beam are made and communicated back to the earth observatory, where image distortion corrections will be made. Such laser probe beams are known as *laser guide star* (LGS) [30–34]. Similar methods may also be employed in FSO systems [35].

In conclusion, the use of adaptive optics in FSO systems and networks creates a narrower beam, increases the link range to beyond 5 Km, reduces scintillation effects, improves optical signal performance, and increases the beam-to-fiber coupling efficiency allowing for all-optical transparent relay nodes and also more WDM channels to be employed in FSO communication links. However, the use of adaptive optics in FSO systems adds to its capital cost and to its operating cost, and therefore an engineering judgment should be made when AdOs are considered.

2.9 LASER SAFETY

Laser safety is an important issue and is concerned with the potential exposure of the eye to the laser beam, because of its ability to focus light on the retina and potentially burn a hole onto it. It is also concerned with the potential exposure to skin. In general, any laser that is "eye-safe" is also "skin-safe."

However, the specific wavelength of light is important in safety because only wavelengths in the range 0.4 and 1.4 μm are transmitted through the cornea of the eye; other wavelengths are absorbed by it. Even so, the power of the laser beam and the time of exposure are important parameters, as well as the spectral range, infrared (IR) or ultraviolet (UV) [36]; in general, the damage threshold is higher for UV than IR.

Lasers are classified in four classes and subclasses, Class 1 is the least powerful and Class 4 the most powerful (see Chapter 1 for classification of lasers).

Most countries have developed or adopted laser safety standards applicable to all products that employ lasers, from a simple laser pointer to most powerful laser that cuts steel. Standards address both, the safety of lasers for various power densities (and not their wavelength) as well as the safe use of lasers; they outline installation compliance requirements based on emitted power density, define specific hazardous zones in front of the transmit aperture, and define installation rules or restriction of certain high-power laser systems in areas that are easily accessible to the public. In addition, they address safety control, the use of protective (eye and skin) equipment, warning labels and signs, automatic shut-off, training, maintenance and service.

Among the standards organizations are (alphabetically):

- American National Standards Institute (ANSI) [37]
- Center for Devices and Radiological Health (CDRH)
- European Committee for Electrotechnical Standardization (CENELEC)
- International Electrotechnical Commission (IEC) [38]
- Laser Institute of America (LIA)

Among the standards that describe recommendations for laser safety are:

- The ITU-T recommendation G.664 provides optical safety procedures.
- The American National Standard Institute (ANSI Z136.1-2000) provides maximum permissible exposure (MPE) limits and exposure durations, from 100 fsec to 8 hours.

- Similarly, the American Conference of Governmental Industrial Hygienists (ACGIH) provides the threshold limit values (TLV) and biological exposure indices (BEI).
- Both ANSI and ACGIH have become the basis for the U.S. Federal Product Performance Standard (21 CFR 1040).

It is important to identify that an FSO node may be classified as Class 1 or 1M, although the actual laser device in the FSO equipment is Class 3B (a relatively high power laser). This is so because the equipment is classified by the output power density before the beam enters the space; for example, a large-diameter lens spreads out the radiation of the laser device so that the power density launched at the aperture of the node is equivalent to Class 1. In addition, Class 1M laser systems that operate at 1550 nm are allowed to transmit approximately 55 times more power than a system operating at 850 nm (shorter IR), for the same size aperture lens. In optical communications, Class 1 and 1M are the important ones; Class 1 systems can be installed in any application without restriction, whereas Class 1M systems should be installed only in locations where the unsafe use of optical aids is prevented.

2.10 NODE HOUSING AND MOUNTING

The physical design of an FSO node entails two parts, the outside housing of the FSO transceiver and its mount (the outside plant or OSP), Figure 2.13, and the router (including fiber-optics and switch) that are in an indoor cabinet (the inside plant or ISP). Between the two, the FSO housing is most challenging because it contains sensitive optical and optoelectronic components, which are continuously exposed to the atmospheric elements, wind, rain, frost, adverse temperature variations (from very high to

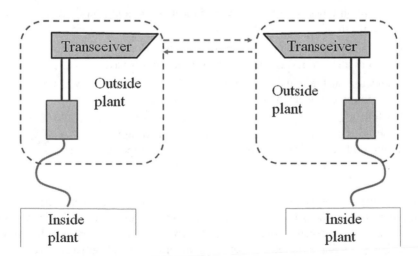

Figure 2.13. Outside and Inside plant of an FSO link.

sub-zero), solar glare, sand or dust-storm, and more. Typically, an outdoors FSO housing needs to operate from −20° to +50° Celcius. Between the FSO housing (indoors or outdoors) and the router, connectivity is provided by a fiber optic cable (single mode or multi-mode, depending on distance and data rates), and untwisted copper pair category 5 with RJ-45 connector for SNMP applications or for node provisioning. An additional cable supplies power to optoelectronics, electronics, heat-stabilizers and auto-tracking mechanism in the housings.

Despite this, the node must operate with a min-max acceptable performance range, if possible at all times, and be compact, relatively lightweight, corrosive free, waterproof, easily serviceable, and have low power consumption.

- Frost and sub-zero temperature is combated with automatic heating elements within the main housing.
- High heat is addressed with a second cover over the main housing that allows for ventilation between the cover and the main housing. In addition, passive or active heat exchange may be deployed.
- Solar glare is addressed by physical design and node orientation so that the transceiver is not directed to sunrise or sunset directly, nor to sun that is deflected by reflecting surfaces.
- Wind pressure is addressed with aerodynamic physical design so that wind exerts minimal pressure to housing and causes no vibrations because of wind turbulence.
- Sand and dust-storms are abrasive to housing surface and particularly to the lens of the housing. Non-abrasive coatings on the housing and specially treated glass minimize the abrasive action of sand and dust-storm.
- Waterproofing is addressed with physical design that assures an environmentally-sealed (airtight and humidity-free) environment within the main housing.
- Light-weightiness is addressed with light metallic and non-corrosive materials, and thin lenses.
- The housing mount should be anti-corrosive and relatively short so that under wind pressure it does not bend to add to building sway.
- Connectors at the housing should be easily accessible and well protected from atmospheric elements, and also tamper-proofed.
- The housing should be modular so that it is easily serviceable; troubleshooting software and hardware incorporated in the node design should be able to identify which module is at fault. Moreover, module or component redundancy, wherever appropriate, is desirable.

FSO housings positioned indoors and behind a window are much simpler. Clearly, the housing of such nodes is not exposed to atmospheric elements and it enjoys a controlled and a relatively secure environment. However, besides the removed functionality associated with being indoors (do not need to have special anticorrosive and anti-abrasive materials, heat stabilizers, and covers), the housing still must be solar glare free, lightweight, modular, and easily serviceable. However, an indoors node has limited applicability because of directional limitations of the beam.

REFERENCES

1. S.V. Kartalopoulos, *DWDM: Networks, Devices and Technology*, IEEE/Wiley, 2003.
2. K. Iga, "Surface-Emitting Laser—Its Birth and Generation of New Optoelectronics Field", *IEEE Journal on Selected Topics in Quantum Electronics*, vol. 6, no. 6, pp. 1201–1215, Nov/Dec 2000.
3. J.S. Harris, "Tunable Long-Wavelength Vertical-Cavity Lasers: The Engine of Next Generation Optical Networks?" *IEEE Journal on Selected Topics in Quantum Electronics*, vol. 6, no. 6, pp. 1145–1160, Nov/Dec 2000.
4. C.J. Chang-Hasnain, "Tunable VCSEL", *IEEE Journal on Selected Topics in Quantum Electronics*, vol. 6, no. 6, pp. 978–987, Nov/Dec 2000.
5. E. Desurvire, *Erbium-doped fiber amplifiers*, Wiley, New York, 1994.
6. S.V. Kartalopoulos, *DWDM: Networks, Devices and Technology*, IEEE/Wiley, 2003.
7. E. Ackerman, S. Wanuga, D. Kasemset, A. Daryoush, and N. Samant, "Maximum dynamic range operation of a microwave external modulation fiber-optic link", *IEEE Trans. Microwave Theory Technology*, vol. 41, pp. 1299–1306, August 1993.
8. U. Cummings and W. Bridges, "Bandwidth of linearized electro-optic modulators", *Journal of Lightwave Technology*, vol. 16, pp. 1482–1490, August 1998.
9. S.V. Kartalopoulos, *Introduction to DWDM Technology: Data in a Rainbow*, IEEE-Press, 2000.
10. ITU-T Recommendation G.661, "Definition and test methods for the relevant generic parameters of optical fiber amplifiers", November 1996.
11. ITU-T Recommendation G.662, "Generic characteristics of optical fiber amplifier devices and sub-systems", July 1995.
12. ITU-T Recommendation G.663, "Application related aspects of optical fiber amplifier devices and sub-systems", Oct 1996.
13. S.V. Kartalopoulos, "A Global Multi Satellite Network", ICC'97, Montreal, Canada, 1997, pp. 699–698. Also in patent # 5,602,838, issued 2/11/1997.
14. S. Lee, J.W. Alexander, and M. Jeganathan, "Pointing and Tracking Subsystem Design for Optical Communications Link between the International Space Station and Ground", *Proceedings of SPIE*, vol. 3932, pp.150–157, 2000.
15. E.J. Korevaar, et al., "Horizontal-link performance of the STRV-2 lasercom experiment ground terminals", *Proceedings of SPIE*, vol. 3615, pp.12–22, 1999.
16. M. Reyes, et al., "Design and performance of the ESA Optical Ground Station", *SPIE Proceedings*, vol. 4635, pp. 248, 2002.
17. A. Alonso, M. Reyes, and Z. Sodnik, "Performance of satellite-to-ground communications link between ARTEMIS and the Optical Ground Station", *SPIE*, vol. 44, pp. 5160, 2003, SPIE MASP03.
18. T.-H. Ho, S.D. Milner, and C.C. Davis, "Fully optical real-time pointing, acquisition, and tracking system for free space optical link," *SPIE, Free-Space Laser Communication Technologies XVII*, G. Stephen Mecherle, Ed., vol. 5712, pp. 81–92, 2005.

19. S.V. Kartalopoulos, "Signal Processing and Implementation of Motion Detection Neurons in Optical Pathways", Globecom'90, San Diego, December 2–5, 1990, pp. 1361–1365.

20. V.S. Hsu, J.M. Kahn, and K.S.J. Pister, "MEMS Corner Cube Retroreflectors for Free-Space Optical Communications", University of California, Berkeley, CA, November 4, 1999, 53 pages, http://www-ee.stanford.edu/~jmk/pubs/hsu.ms.11.99.pdf, 9/2010.

21. R.L. Austin and R.J. Schultz, "Guide to retroreflection Safety Principles and Retroreflective Measurements", November, 2006, RoadVista, San Diego, CA, http://www.atssa.com/galleries/default-file/RetroreflectionGuide-ATSSA.pdf, 9/2010.

22. K. Strehl, "Aplanatische und fehlerhafte Abbildung im Fernrohr", *Zeitschrift für Instrumentenkunde*, vol. 15, pp. 362–370, October 1895.

23. K. Strehl, "Über Luftschlieren und Zonenfehler", *Zeitschrift für Instrumentenkunde*, vol. 22, pp. 213–217, July 1902.

24. J. Liang, D.R. Williams, and D.T. Miller, "Supernormal vision and high-resolution retinal imaging through adaptive optics", *Journal of Optical Society of America*, vol. 14, pp. 2884–2892, 1997.

25. S. Chandrasekhar, *Liquid Crystals*, Cambridge University Press, 1994.

26. P.G. de Gennes and J. Prost, *The Physics of Liquid Crystals*, Oxford, Clarendon Press, 1993.

27. H. Hammati, "Overview: Free-Space Optical Communications at JPL/NASA", March 2003.

28. H. Hammati, "Overview of LserCommunication Research at JPL", *Proc. SPIE*, vol. 4273, The Search for Extraterrestrial Intelligence (SETI) in the Optical Spectrum III, August 2001.

29. J.R. Lesh, L.D. Deutsch, and W.J. Weber, "A Plan for the Development and Demonstration of Optical Communications for Deep Space," Proceedings of Optical Communications II Conference at Lasers'91, Munich, Germany, June 10–12, 1991.

30. A. Primmerman, et al., Compensation of atmospheric optical distortion using a synthetic beacon, *Nature*, vol. 353, pp. 141–143, 1991.

31. Laser Guide Star System on ESO's VLT Starts Regular Science Operations, http://www.eso.org/public/news/eso0727/, retrieved September 2010.

32. P.L. Wizinowich, et al., "The W. M. Keck observatory laser guide star adaptive optics system: overview", *Publications of the Astronomical Society of the Pacific, 2006 Conference*, vol. 118, pp. 297–309.

33. N. Hubin and L. Noethe, "Active Optics, Adaptive Optics, and Laser Guide Stars", *Science*, vol. 262, no. 5138, pp. 1390–1394, 26 November 1993.

34. C.C. Chien and J.R. Lesh, "Application of Laser Guide Star Technology to Space-to-Ground Optical Communications Systems", presented at the Laser guide star Adaptive Optics Workshop, Phillips Laboratory, Kirtland AFB, Albuquerque, NM, March 10–12, 1992.

35. "AOptix Technologies Introduces AO=Based FSO Communications Products", June 2005, http://www.adaptiveoptics.org/News_0605_1.html. Retrieved September 2010.

36. W.L. Wolfe, *Introduction to Infrared System Design*, SPIE Press, 1997.

37. American National Standards Institute (ANSI), "Safe Use of Lasers", ANSI Z136.1-2000, 2000.

38. IEC 60825–1, Amendment 2 Laser Power Levels.

3

POINT-TO-POINT FSO SYSTEMS

3.1 INTRODUCTION

Free Space Optical technology is applicable to distances that range from few hundred meters to few kilometers (in terrestrial applications), and to thousands of kilometers (in earth to satellite and in intersatellite applications); FSO may also be used in (highly specialized) deep space applications. In addition, FSO supports prevalent and standardized protocols, such as SONET/SDH, Ethernet, and others. Because the link distance is not the same in all applications, the laser power does not need to be the same; shorter links require weaker lasers than longer links. However, the attenuation of the medium (the atmosphere) is not constant but it changes dramatically (compare a clear day with few dB attenuation per kilometer to a dense foggy day with hundreds of dB attenuation per kilometer). Therefore, the laser power needs to be self-adaptive for link reliability and signal performance communications; that is, an automatic adaptive optical power control mechanism should exist that adjusts the laser power autonomously to maintain the signal performance at the expected level. Only in severe attenuation cases (very dense fog or hail), the performance level is beyond control and another parallel mechanism is necessary, such as RF, to maintain the link operational.

Free Space Optical Networks for Ultra-Broad Band Services, First Edition. Stamatios V. Kartalopoulos.
© 2011 Institute of Electrical and Electronics Engineers. Published 2011 by John Wiley & Sons, Inc.

In the previous two chapters, we examined the medium through which a laser beam travels the atmosphere, the various components that make an FSO transceiver, how two transceivers establish and maintain connectivity (acquisition, pointing and tracking), and issues related with housing and mounting an FSO transceiver.

In this chapter, we start with FSO topologies and we particularly examine the simplest and first topology used, the point-to-point (PtP).

As already emphasized, two transceivers require line of sight (LoS) to establish a link. This means that one transceiver, regardless of where it is positioned, must be able to "see" the other transceiver; that is, in terrestrial applications, the other transceiver cannot be beyond the horizon or behind obstructions.

Typical PtP terrestrial applications consist of two transceivers that are mounted on top of two buildings to establish a FSO link, whereas in some cases, one or both transceivers are mounted behind a window. Currently, laser beams operate at the popular and well-established in fiber-optic communications 1550 nm wavelength, although wavelengths in the 780–920 nm range have been used successfully. The 800 nm wavelength can employ low cost lasers and receivers, whereas the 1550 nm wavelength has the lowest attenuation in silica-based materials; this is important when the laser beam is positioned behind a window, as in window-to-window (WtW) applications. However, because FSO requires line of sight, and although WtW is preferred for reasons that will become clearer in this chapter (no weather hardening necessary), FSO is achieved from window to roof-top, or from roof-top to roof-top.

In any case, PtP FSO requires two transceivers, and switches or routers to connect with the network. As shown Figure 3.1, the two transceivers are connected with the electronic routers via cables, whereby the routers are located in cabinets within the building, from where they are connected with the public or private network.

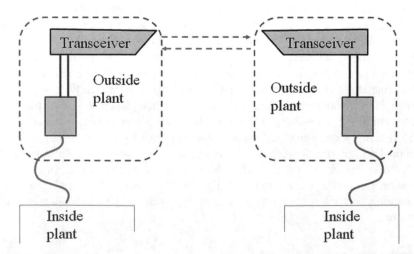

Figure 3.1. Point to Point topology; one (or both) transceiver is connected with the public data network.

The PtP FSO transceiver node is the simplest with regard to design, acquisition, pointing and auto-tracking, as compared with more complex topologies that are presented in subsequent chapters. For example, in PtP topology, the typical laser power, at about $\lambda = 800$ nm, is less than 15 mW (Class I) for link length up to 2 km, a tracking range of approximately ±1.2° in the vertical and in the horizontal direction and they operate in the temperature range −20° to 50°C; acquisition and pointing is accomplished by telescope. More powerful nodes have a range of 6 km for data rates up to OC3/STM-1 (~155 Mb/s) or Fast Ethernet (~125 Mb/s) [1, 2]. Longer links are also possible but at lower data rate, or shorter links at high data rate. Currently, 10 Gb/s has been deployed for moderate link lengths (<1 km).

Because of simplicity, coupled with the fact that laser light is in the unlicensed electromagnetic spectrum, safe low laser power, and no need for cumbersome permits or street trenching, the PtP topology was deployed first and it demonstrated that service was quickly offered at data rates commensurate with fiber-optic networks, and that it was tolerant to moderate atmospheric phenomena (snow, fog, wind, scintillation, beam divergence).

3.2 SIMPLE PtP DESIGN

A point-to-point transceiver consists at minimum of a laser transmitter and a photodetector receiver within a waterproof housing that can be rigidly mounted on a pole or on a wall; to this, an auto-tracking mechanism may be added as well as provisions for mounting a telescope on it; when alignment has been accomplished, the telescope may be removed. In addition, the housing includes a protected compartment with connectors where all external cables, communications and power, are connected. Now, although the housing may seem simplistic and trivial, its implementation is not. Most nodes have double, quadruple or more laser beams, whereas the photodetector has a lens to increase aperture. Thus, the optics, optoelectronics and electromechanical devices within the housing become more complicated.

If the housing is mounted outdoors, it is exposed to wind, rain, fog, hail, snow, etc., and it must be able to operate at extreme temperatures (typically, from about −25°C to about 70°C). In addition, to assure that frost will not cover the optical windows for the lasers and the lens, antifogging devices need to be added, as well as devices that remove or add heat accordingly to maintain the operational temperature in the housing and within the specified temperature range.

If the housing is mounted indoors, the physical design is simplified but not the optoelectronics, adaptive laser power control, and electromechanical functionality for auto-tracking.

Typically, a practical FSO node, for indoors or outdoors, consists of three parts; the housing with all components already mentioned (some call it "the head"), an interconnection unit (typically a small waterproof metallic box), and a switch or a router. The interconnection unit is where cables from the "head" and cables from the switch are interconnected; this allows for easy installation and service of the head and of the switch, and it may also include a power supply unit.

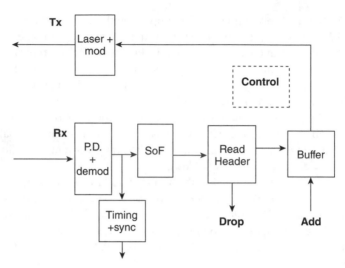

<u>Figure 3.2.</u> Major functional blocks of a simple PtP transceiver. The functional blocks for testing, acquisition, pointing and tracking are not shown.

The overall system should also include monitoring facilities and visual indicators for signal on/off, power on/off, link status, signal quality and excessive bit error indication, as well as management and control capabilities.

3.2.1 Simple PtP Transceiver Design

The functionality of a PtP transceiver is shown in the block diagram of Figure 3.2. In it, it is assumed that information is "packetized" or "framed" according to a standard protocol, such as SONET/SDH, E3/DS3, Ethernet, or others. As such, the received optical data is detected by a photodetector-demodulator and it is converted to binary electronic data, while clock (timing) is also recovered and the start of frame/packet in the data stream is found for synchronization. Then, the header is read to determine the destination of packet, length of packet; possibly, error control may also be performed.

Because in this simple case, the node that has received data is also the destination, then data is "dropped", and new data is "added", buffered and it is directed to the transmitter. Notice that the functions shown in the diagram of Figure 3.2 are typically separated in physical layer functions (laser/photodetector and optics, modulation/demodulation) that are contained in the head of the FSO transceiver, and higher layer functions (synchronization and timing, switching, error control, etc.) that are contained in the router or in the switch; however, as functionality integration is continuously increasing, the head will eventually include more functionality (physical layer +) to simplify design and to avoid unnecessary router delays. This will become more evident in subsequent sections. Additional functions, not shown in the diagram of Figure 3.2, are signal performance monitoring, automatic laser power adaptation, testing, and auto-tracking.

Note: Depending on design objectives, many transceiver heads have redundant lasers to meet different needs and applications; depending on manufacturer and head design, there may be from one to eight laser transmitters.

Simple PtPs always terminate (drop and add) data at the transceivers of the link. However, this design allows for a linear PtP link with intermediate add-drop nodes that becomes clearer in the next section.

3.2.2 Effect of Wind on LoS

All buildings, on which strong winds exert pressure, bend and vibrate. However, the amount of dynamic or static wind-induced lateral displacement and the vibration modes and frequencies at roof-top from the position at rest depends on many factors. Among them is the height of the building, whether the building is stand-alone or it "leans" against other buildings on its sides, on type of materials, elasticity, structural design, on the damping factor, on architectural and aerodynamic design, on topography or building location (on a hill or in a valley), on wind dynamics and wind direction, on temperature and humidity, and more [3, 4].

The wind-induced displacement consists of three parts: a static response due to a 10-minute average wind velocity, a static response due to wind turbulence, and a dynamic response caused by resonance due to wind gust. In general, the amount of displacement at roof-top increases as a building becomes taller and as the wind becomes stronger. For a tall building, a displacement of 1–2 meters for strong winds may be typical, and also a natural frequency (of building oscillation) of less than 1 Hz.

As a consequence, the building displacement and vibration modes due to wind affect the line of sight of FSO links. Therefore, the location of the FSO head, the head mount and the tracking characteristics become an integral part of the building dynamics under wind load. For example, the wind affects the line of sight of a FSO head mounted on the 90th floor of a steel and glass skyscraper differently than a FSO head mounted on the 3rd floor of a low stone and brick building. Moreover, the mount of the head should be as short and as rigid as possible so it does not exacerbate the displacement and vibration modes.

3.2.3 Simple PtP Power Budget Estimation

The main objective of estimating the power budget over a link is to assure link availability (a typical availability is in the order of better than 99.99%); that is, the power of the optical signal at the receiver is greater than or equal to its sensitivity, and that the signal arrives at the expected quality and expected performance (BER and OSNR). The link budget establishes the operating parameters of the link, from transmitter to receiver (included) and is estimated in a similar way as the link budget in fiber optic links [5]. In general, the link budget calculation starts from the receiver towards the transmitter.

The power link budget involves all losses and gains on the path of the optical signal, including loss due to beam divergence, loss due to optics, loss due to connectors and loss due to the atmospheric medium. The latter is a variable that varies substantially over time as it depends on atmospheric conditions. Therefore, the link budget should be estimated using a good model for the atmosphere as the propagating medium and also a worst case scenario commensurate with an acceptable performance level at the data rate on the link.

For example:

- transmitter output power
- losses of optical components on the beam path (filters, lens, protective windows)
- Fresnel reflections, or reflection at windows and lens
- detector sensitivity
- detector noise
- length of link
- beam divergence angle
- beam misalignment
- amount of building sway due to wind or temperature
- attenuation due to atmospheric effects
- and more.

Power gain and loss is additive when expressed in dB units, and thus the power budget estimation reduces to a straightforward addition and subtraction. Additionally, a power margin is added to account for unpredictable losses above and beyond the loss due to beam shift with respect to the receiver as a result of wind or building thermal expansion. The margin is typically set to few dB. Thus, the link power budget starts from the receivers and it works its way back to the transmitter, including power margin, and it expressed by the general relationship:

$$\text{Receiver Sensitivity} = \text{Transmitter output power} - \text{Margin} - \Sigma(\text{losses}), \quad (\text{dB}) \qquad 3.1$$

where the sum of losses includes all power losses on the path, from the laser device aperture to the photodetector; this includes lens and protective window losses, filters, actual optical power at the receiver (calculated from the beam divergence, link length and beam profile, and it is expressed as power loss in dB), gain of focusing lens (focusing is expressed as gain instead of loss), and atmospheric losses (a variable that needs to be estimated under worst case).

In general the power budget estimation is in four steps:

1. The *receiver signal level* (RSL) is determined from the photodetector sensitivity that is commensurate with the expected signal performance.
2. Working backwards (from receiver to transmitter), the free-space *link loss* (LL) is estimated, including all relevant losses associated with the medium, the beam, building sway, etc, as well as gain at the receiver that depends on the aperture of the collecting lens, and the added margin.

3. Then, the *effective isotropic radiated power* (EIRP) of the transmitter is estimated.

4. From the estimated EIRP, and knowing additional associated losses at the transmitter, the source (laser) power is calculated.

The EIRP may also be calculated from the specifications of the receiver and the total free-space link loss:

$$\text{EIRP} = \text{RSL} + \text{LL} + \text{Margin (dB)} \qquad 3.2$$

In this case, the optical transmitter power, P_t, is:

$$P_t = \text{EIRP} - G_t \text{ (dB)} \qquad 3.3$$

where G_t is the transmitter gain

Now, consider a source with gain G_t that transmits optical power P_t. The radiated power P_d at a distance d is:

$$P_d = P_t G_t / 4\pi d^2 \qquad 3.4$$

Because of propagation losses and beam divergence, only part of the beam is "seen" by a receiver with aperture A_r.

Now, if the gain of the receiver is

$$G_r = 4\pi A_r / \lambda^2. \qquad 3.5$$

then, the power received by the photodetector (that is, the receiver signal level) is:

$$P_r = P_d A_{\text{eff}} = P_t G_t G_r (\lambda/4\pi d)^2 \qquad 3.6$$

where $A_{\text{eff}} = A_r/A_t$, and A_t is the aperture (if any) of the transmitter, else, $A_{\text{eff}} = A_r$. The term $(\lambda/4\pi d)^2$ represents the decrease of power flow in free-space at distance d and it is expressed in Watts/m^2.

In the aforementioned equation, if the margin and the detector noise components are removed, and if the atmospheric attenuation constant α is modeled to be close to realistic, then the power at the receiver P_r is expressed as:

$$P_r = P_t \cdot \{A_r / (\theta \cdot L)^2\} \cdot e^{(-\alpha L)} \qquad 3.7$$

where P_t is the transmitter power (in dBm), θ is the beam divergence angle (in radians), and L is the link length (in km).

The latter equation instructs that the amount of received power is proportional to the amount of power transmitted and of the area of the collection aperture. However, it is inversely proportional to the square of the product {beam divergence and link range}, and also inversely proportional to the exponential of the product {atmospheric attenuation coefficient and the link range}. Thus, because the link length L for most applications is fixed, the two controllable parameters are the transmitter power and the aperture of the receiver. Because the atmospheric attenuation coefficient is not under our control and because it dominates the performance of FSO systems exponentially, a worst case power budget scenario should be considered as well as a sufficiently large margin, a large receiver aperture, and adaptive power laser methods.

The aperture of the receiver can be sufficiently large by using a large focusing lens, in which case the irradiance gain G_i should be calculated (dBi).

Finally, the operating wavelength should also be known as the loss due to atmospheric effects depends on wavelength (see chapter 1). In certain applications, the link length may also be controllable if a point-to-point with repeaters topology is used. However, one should also remember that the signal noise, noise factor, and signal-to-noise ratio are cumulative over the total path.

3.2.4   Modeling Atmospheric Loss

Although most of the afore-enumerated loss components are fairly well-known and well-estimated, atmospheric loss is complex and is estimated using probabilistic and statistical models. However, the accuracy of the model is critical because the atmosphere is a major loss contributor, it changes dynamically, and it has a large dynamic range, from a fraction of a dB to more than 300 dB per kilometer (in moderate fog, the attenuation is 100 dB/km).

Atmospheric turbulence due to temperature variations and its impact on FSO transmission has been studied and analyzed [6–9], and models have been proposed, such as the Gamma-Gamma distribution for small-scale and large-scale atmospheric fluctuations due to turbulence [10].

A more general model for the atmosphere is the *Digital Imaging and Remote Sensing Image Generation* (DIRSIG) that was developed at the Rochester Institute of Technology [11]. The physics implemented by this model produce simulated realistic imagery in the region from visible through thermal infrared and it is designed to produce broad-band, multi-spectral and hyper-spectral imagery through the integration of a suite of radiation propagation sub models based on fundamental physics, chemistry and mathematical theories. Such theories include, but not limited, light-mater interaction, Fresnel theory, thermal conductivity, density, radiation absorption factors, radiation and convective loadings, flow of energy, and so on.

The DIRSIG models the atmosphere with acceptable accuracy using an atmospheric radiative propagation sub-model known as MODTRAN, which has been used by the United States Air Force [12, 13]. MODTRAN, under different conditions, predicts properties of path radiances, path transmission, sky radiances and surface reaching solar and lunar irradiance for a wide range of wavelengths and spectral resolutions.

3.3   POINT-TO-POINT WITH TRANSPONDER NODES

Engineering an FSO simple point-to-point link may become more cumbersome if the link length is beyond the acceptable limits, as defined by power budget and link availability with expected signal performance, it is rather straightforward.

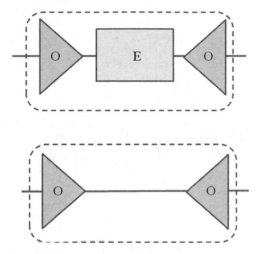

Figure 3.3. A. O-E-O and, B. O-O transponders.

However, in some applications the link length is much longer than what the power budget defines, or there is not a clear line of sight between the two end nodes. In this case, the overall link is subdivided and between the two end nodes one or more intermediate nodes are placed, which act as transponders or relays. These transponders have two sets of transceivers, and they do not have add-drop or any signal integrity restoration capability. However, noise generated by each transponder is cumulative and therefore the number of transponders should be limited.

There are two types of transponders: optical-to-electrical-to-optical (O-E-O), and optical-to-optical (O-O).

The O-E-O transponder may also act as a 3R repeater; that is, it performs signal reshaping, retiming, and reconstitution or gain; O-E-Os are more complex and more expensive, Figure 3.3. Because the signal is converted to electronic, an O-E-O node allows for add-drop functionality, in addition to simple optical relay or transponder.

The O-O transponder, or optical relay, is technologically more attractive because it performs direct optical-to-optical amplification using optical amplifiers (doped fiber-based (EDFA) or semiconductor optical amplifiers (SOA)) thus acting as an all-optical relay.

3.3.1 PtP Transceiver Design with Add-Drop

Point-to-Point links may require an intermediate add-drop node, which receives data and it either drops it at the site or it passes it to the next node. Such links are known as *Point-to-Point with add-drop nodes*. In this case, the intermediate transceivers act either as relays (when they pass data to the next node) or as terminating nodes (when they add and drop data) [14–16].

Although the far end nodes of a Point-to-Point with add-drop nodes are designed as in the previous section, the intermediate nodes need to have two transceivers as well

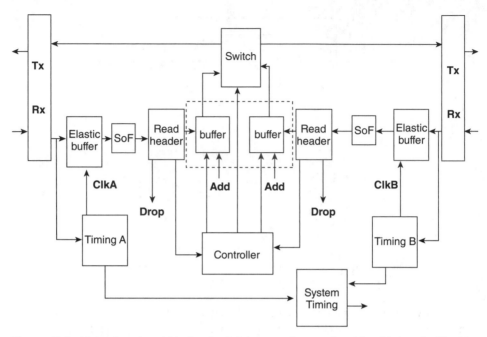

Figure 3.4. Major functional blocks of a PtP intermediate node with add-drop facility. The functional blocks for testing, acquisition, pointing and tracking are not shown.

as a data switch facility for the add-drop capability. Figure 3.4 illustrates the functional block diagram of the intermediate node explicitly. Note that this block diagram is amenable to circuit integration in which case several blocks can be consolidated (buffers, header readers, timing, and others) to yield smaller and cost-effective circuitry. In the diagram of Figure 3.4, testing, acquisition, pointing and tracking mechanisms are not shown. Commercial systems designed and manufactured by various companies to meet different service needs, and at different cost points, are many; a search on the world wide web (WWW) will be very fruitful.

3.3.2   Power Budget for PtP with Add-Drop Nodes

As in the previous case, the main objective of estimating the power budget over a link is to assure link availability with the expected signal quality and performance (BER and OSNR). However, in this case the link is partitioned in smaller link segments, whereby each intermediate node consists of two transceivers and additional paths for adding and dropping traffic, Figure 3.5. Thus, the link budget is calculated from node to node, whereby each intermediate node is assumed to act as a 3R optical-to electrical-to optical (O-E-O) repeater with gain (that is, it performs signal reshaping, retiming, and reconstitution or gain), where the gain counterbalances losses in the previous segment.

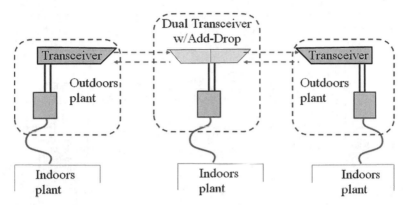

Figure 3.5. The power budget in a PtP link with add-drop nodes is calculated for each segment between two successive nodes. An add-drop node acts as a repeater.

A more technologically attractive repeater performs directly optical-to-optical using optical amplifiers (EDFA or SOA) at each repeater node, thus acting as an all-optical repeater, although adding and dropping traffic in the optical regime is a challenge for a FSO technology that is supposed to be cost-efficient and inexpensive. However, if advances in the optical regime allows for an all optical add-drop, then, the link budget is performed over the total link, whereby the gain and loss at intermediate nodes from the added and dropped paths should be carefully examined. Moreover, the amplifier noise as well as noise from the added signal should be taken into account in the power noise budget, since noise is cumulative. Note that if the FSO beam consists of many wavelengths (that is, WDM), then optically adding and dropping is a well-established optical technology that has already been deployed in fiber metro ring networks.

3.4 HYBRID FSO AND RF

The FSO link availability is expected to be better than 99.99% and at the expected signal performance. Now, for FSO that depends on atmospheric phenomena this is a very ambitious target. Depending on geography, some areas have more often than others dense fog, hail, snow, rain, sand-storms, and so on. Thus, when dense fog settles in an area, few hours of service outage ruins the prescribed expected service availability.

For service to be available 99.99%, a back-up communication strategy is needed; for terrestrial applications, this strategy encompasses a symbiotic hybrid technology that consists of a primary FSO link working in parallel with a secondary broadband radio frequency (RF) link; as soon as the FSO link becomes inoperable or its performance drops below the expected level, the RF link becomes operable to continue uninterrupted service, Figure 3.6. The RF frequency uses a high gain (high directionality) antenna.

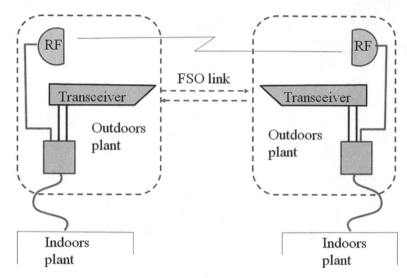

Figure 3.6. Hybrid FSO with RF link.

However, for the hybrid link to work efficiently, a fast responding protocol based on a closed loop should exist. For example, although the optical link may be operable, the RF link may be sending performance parameters periodically, so that as soon as the performance is degraded the RF link becomes primary; the RF link is secondary because at long-wavelengths only a fraction of the FSO data rates can be supported, although it propagates through fog and over long distances. Broadband RF technology is mature, inexpensive, and has been very effectively used by Telcos.

For a moderate link length up to 1 km, millimeter waves in the unlicensed spectrum may also be employed to support data rates comparable to FSO links; such frequency is 60 GHz, but it is more susceptible to rain than fog (the opposite of FSO links).

Additionally, environmental monitoring mechanisms may be included at each node, which based on an atmospheric model predict imminent outages due to fog, rain, and so on. However, these mechanisms add to cost and node complexity and their cost needs to be justified by the particular application.

Needless to say that in either case, accurate auto-tracking is necessary to avoid misinterpretation of signal degradation due to wind displacement.

For earth to space applications, the back-up frequencies are in the GHz range using high gain (paraboloid) antennas to overcome Oxygen absorption.

3.5   FSO POINT-TO-MULTIPOINT

The FSO point-to-multipoint topology may be considered in access networks, Figure 3.7. This technology becomes complex in the downstream direction than the upstream for the reason that many beams need to be employed, each beam for each end node in

Figure 3.7. Concept of FSO point-to-multipoint topology.

the multipoint topology. Two cases are envisioned in this topology: the first employs a broadcasting beam that is spread over a wide angle to connect with all multipoint end nodes, and the second employs multiple beams each directed to each end node. Clearly, the first is simpler and more economical than the second.

The FSO point-to-multipoint topology, as described, has not yet been deployed. However, a more viable topology uses a point-to-point FSO link up to a cluster of nodes, whereby the optical signal is converted and broadcasted to multiple nodes over RF using one of the newer protocols, such as the Mobile Worldwide Interoperability for Microwave Access (WiMAX), which supports downstream data rates up to 37 Mb/s and upstream up to 10 Mb/s [17].

3.6 FSO POINT-TO-MOBILE

This part describes a PtP FSO topology, which consists of a main stationary node and one or more mobile nodes. This topology is suitable in communications applicable between an approaching aircraft and the airport tower, between a ship and moving vehicles on the shore (ship-to-shore), between a stationary earth-station and low-earth orbiting satellites (LEOS), and so on. However, these methods are as good as the self-tracking system in the nodes, as well as the line of sight capability.

In the first application, a large amount of information related to data associated with landing needs to be transmitted in a very short time because of the fast approaching aircraft; for example, the approaching speed for landing of a Boeing 747 is ~180 mph (~290 km/h), or 3 mi. (~5 km) per second. If an airplane is 10 km away from touchdown and corrections must be made (direction, angles, speed, etc), the amount of time available for data communication, response time (corrective action), and verification before touchdown is very short. In this case, a laser beam that transports necessary data in a small fraction of a second saves time that may be critical in certain cases.

In the case of ship-to-shore, a laser beam provides higher communication security because the RF technology commonly used is omni-directional and as such it is also received by foe receivers. Because of the narrowness of the laser beam, it is more unlikely that communications security can be compromised. The amount of data transported over the beam may be large (Gb/s) or moderate (~100 Mb/s), which in any case is sufficient to transport all types of data (voice, video, image, files). However, line of sight is paramount and therefore this technology very much depends on geography. If the geography does not allow for direct line of sight, it may also be possible to have an aircraft hovering over the shore with an a FSO transponder on board; obviously, this case requires sophisticated autonomous tracking systems on the ship and on the aircraft.

In addition to all-optical FSO point-to-mobile networks, there are proposals for hybrid FSO and RF mobile ad-hoc networks (MANET). Integrating a hybrid FSO and MANET network however poses its own challenges particularly in fast handover, mobile node addition and deletion (as part of the ad-hoc networks), maintaining LoS and bandwidth requested by each node in the ad-hoc network, management, testing, cost-efficiency, and obviously dynamic and autonomous tracking.

REFERENCES

1. S.V. Kartalopoulos, *Understanding SONET/SDH and ATM*, IEEE/Wiley, 1999.
2. S.V. Kartalopoulos, *Next Generation SONET/SDH: Voice and Data*, IEEE/Wiley, 2004.
3. ASCE 7-02 Standard, *Minimum Design Loads for Buildings and Other Structures*, SEI/ASCE, 2002.
4. Council on Tall Buildings and Urban Habitat, "Tall Buildings in Numbers. Tall Buildings in the World: Past, Present and Future," *CTBUH Journal*, vol. 2, pp. 40–41, 2008.
5. S.V. Kartalopoulos, *DWDM: Networks, Devices and Technology*, IEEE/Wiley, 2003.
6. X. Zhu and J. M. Kahn, "Free-space optical communication through atmospheric turbulence channels," *IEEE Trans. Communications*, vol. 50, pp. 1293–1300, August, 2002.
7. X. Zhu and J. M. Kahn, "Performance bounds for coded free-space optical communications through atmospheric turbulence channels," *IEEE Trans. Communications*, vol. 51, pp. 1233–1239, August, 2003.
8. M. Uysal, S. M. Navidpour, and J. T. Li, "Error rate performance of coded free-space optical links over strong turbulence channels," *IEEE Communications Leters*, vol. 8, pp. 635–637, October, 2004.
9. M. Uysal and J. T. Li, "Error rate performance of coded free-space optical links over gamma-gamma turbulence channels," in Proc. of IEEE International Communications Conference (ICC'04), Paris, France, pp. 3331–3335, June, 2004.
10. M. A. Al-Habash, L. C. Andrews, and R. L. Philips, "Mathematical model for the irradiance probability density function of a laser beam propagating through turbulent media," *Optical Engineering*, vol. 40, pp. 1554–1562, August, 2001.
11. http://www.dirsig.org/docs/manual/.

12. http://www.rese.ch/pdf/MODO_Manual.pdf. Retrieved 23 January, 2011.
13. Berk et al., 1999, MODTRAN4 User's Manual, Air Force Research Laboratory.
14. J. Akella, M. Yuksel, and S. Kalyanaraman, "Error analysis of multi-hop free-space-optical communication," *IEEE International Conference on Communications (ICC)*, vol. 3, pp. 1777–1781, 2005.
15. V. Gambiroza, B. Sadeghi, and E. W. Knightly, "End-to-end performance and fairness in multihop wireless backhaul networks," *ACM Mobicom*, pp. 287–291, 2004.
16. J. F. Labourdette and A. Acampora, "Logically rearrageable multi-hop lightwave networks," *IEEE Transactions on Communications*, vol. 39, no. 8, pp. 1223–1230, 1991.
17. IEEE 802.16e, "IEEE Standard for Local and metropolitan area networks Part 16: Air Interface for Fixed and Mobile Broadband Wireless Access Systems," 2005.

4

RING FSO SYSTEMS

4.1 INTRODUCTION

Free Space Optical technology has been successfully deployed in many applications: terrestrial, space, commercial, and military. The laser beams used in FSO applications are in the THz range (1.55 μm) and thus they are capable of supporting a huge data rate (up to several Gb/s) and therefore many more end-users, compared with RF or Microwave technology. Current cost-effective applications include:

- Voice, Data and Video
- Enterprise
- Hard to reach areas
- High-bandwidth network on limited time
- Temporary service activation
- Disaster avoidance
- Disaster recovery
- Surveillance
- Last/First mile access

Free Space Optical Networks for Ultra-Broad Band Services, First Edition. Stamatios V. Kartalopoulos.
© 2011 Institute of Electrical and Electronics Engineers. Published 2011 by John Wiley & Sons, Inc.

- Telco bypass
- Local area networks (LAN) and Metropolitan access networks (MAN)
- Point-to-point, Point-to-point with add-drop, Ring and Mesh topologies, Figure 4.1.

FSO point-to-point (PtP) topology is applicable to distances that range from few hundred meters to few kilometers (in terrestrial applications), and to thousands of kilometers (in earth to satellite and in intersatellite applications). FSO PtP may also be used in highly specialized deep-space applications.

In terrestrial applications for which the link exceeds the permissible link length, FSO relays may be used or FSO add-drop nodes as described in the previous Chapter. In such case, FSO nodes with add-drop capability consist of more than one transceiver and they may be applied to metropolitan ring networks in a ring topology. In fact, because the FSO links are full-duplex (bidirectional), the constructed ring is of the type two counter-rotating rings metropolitan area network (MAN), Figure 4.2.

The ring and mesh topologies are not new in communications and their failure mechanisms, and service protection schemes or countermeasures have been studied extensively in fiber-optic local area networks (LAN) and metro area networks (MAN) applications [1–3]. In fiber-optic LAN and MAN ring network applications, there are three ring configurations: single ring, dual ring and quad ring.

4.2 RING TOPOLOGIES AND SERVICE PROTECTION

Single ring topology with unidirectional links is most vulnerable to failures, because a single failure stops the flow of data around the ring, Figure 4.3.

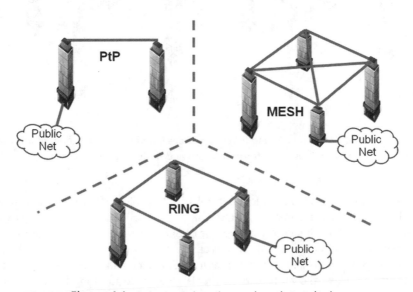

Figure 4.1. Point-to-Point, ring and mesh topologies.

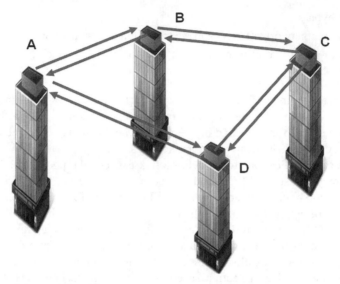

Figure 4.2. Dual full-duplex FSOs can form a two counter-rotating rings LAN.

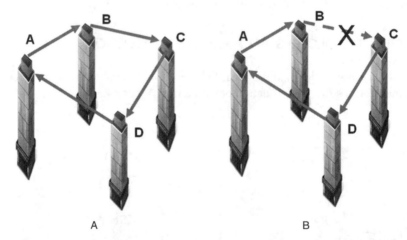

Figure 4.3. A single fault on the ring (between nodes B and C) stops the flow of information in simple unidirectional rings. A. Before the fault, and B. during the fault.

The ring topology with bidirectional links is also vulnerable to failures, but using traffic loop back mechanisms the failure link is isolated and service continues to be provided, Figure 4.4. This case is suitable to FSO ring topology since FSO links are full duplex and thus bidirectional. However, the FSO node design should be able to allow for loop-backs (as shown in the previous Chapter), and to also have monitoring mechanisms to detect the fault, and protocols to activate the loop-back mechanism. Loop-backs may take place at the physical layer at the transceiver head (from the

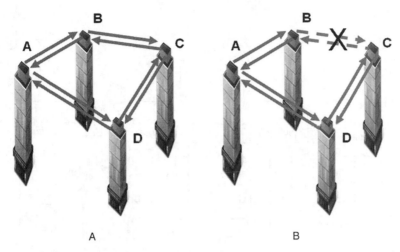

Figure 4.4. Ring topology with bidirectional links form two counter-rotating rings (A); with fault monitoring and data loop-back mechanisms they are capable to isolate a faulty link (B). Loop-backs are performed at node B and C.

photodetector to the laser), or at the MAC layer (in the router memory), Figure 4.5a and 4.5b.

The dual ring topology with unidirectional links is vulnerable to failures, but with counter-rotating traffic and traffic loop-back mechanisms, the failure link is isolated and service continues to be provided. This is also applicable to FSO ring topology since FSO links are full duplex or bidirectional, and thus they easily construct a dual counter-rotating ring. However, FSO nodes should have provisions for fault detection, for looping-back data (as shown in the previous chapter), and also protocols that activate data looping-back.

The quad ring topology is also vulnerable to failures. However, this topology consists of two pairs of counter-rotating rings, and each node has failure detection mechanisms, multiple data loop-back facilities, and more complex protection protocols. Although pointing and auto-tracking is the same as in simple nodes, they are more expensive, and thus they are justified in applications that demand high survivability, service high availability, and network high availability.

4.3   RING NODES WITH ADD-DROP

Figure 4.4, above, illustrates a ring with bidirectional links and with nodes that support add-drop. In the previous chapter we described the FSO node design with two transceivers and with add-drop. These nodes are also applicable to nodes in the bidirectional-link ring topology. The add-drop is a simple two-by-two switch which either passes through traffic or it adds and drops, or it loops back traffic; loop-backs are suitable for testing links and nodes as well as avoiding faulty links, Figure 4.5.

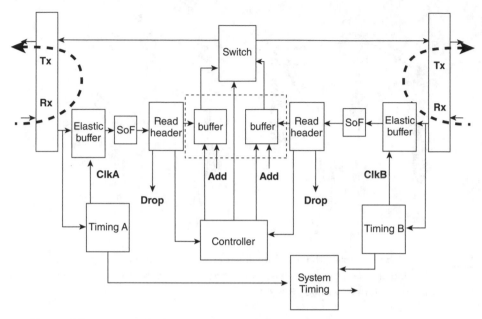

Figure 4.5a. Looping back data at the physical layer (from photodetector to laser).

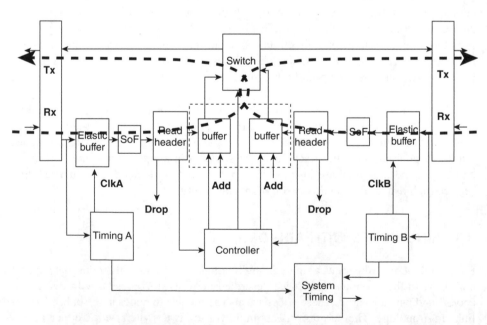

Figure 4.5b. Looping back data at the MAC layer (through the memory buffer).

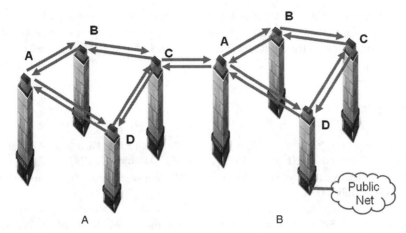

Figure 4.6. A bridge on a FSO ring (nodes C and A) passes data to its adjacent ring.

The add-drop traffic, because of its high bandwidth, supports many end-users simultaneously. Thus, the add-drop path may be connected to several end-terminals.

The ring topology with add-drop is also applicable to geostationary satellite networks. In this case, each satellite comprises a node of the ring and the inter-satellite links (ISL) are with laser beams and/or microwave links. In fact, for better network survivability, each ISL may consist of more than one laser beams (one working and one spare or "protection" beam, known as "1 + 1 protection").

Because satellite networks with ISLs are able to route traffic directly from node to node, they reduce delay and echo that is a nuisance in communications; they improve the quality of signal, and also network protection. Such "network-in-the-sky" was envisioned decades ago but now is a reality.

4.4   CONCATENATED RINGS

In networks, a ring (or a LAN) by itself is not practical. It is practical, however, when a ring (or LAN) is connected with another ring (or LAN). In this case, one of the nodes is designed so that it entails a node of the ring and also a bridge, which is connected with the bridge of another similar ring. This case is also known as peer-to-peer connectivity. Thus, traffic the flows around one ring, is shorted out at the bridge and packets or frames that are destined for the adjacent ring are detected by the bridge and are passed over to the adjacent ring, Figure 4.6.

4.5   RING TO NETWORK CONNECTIVITY

In actual networks, a ring (or a LAN) needs to be connected with the network so that communications connectivity from any node of the ring can be established with any other node in the world, as shown in Figure 4.6. In this case, the node of the ring that connects with the network is capable to map its LAN protocol onto the protocol used by the network, such as for example, Ethernet over SONET/SDH, Ethernet over DS3, and so on [4].

REFERENCES

1. S.V. Kartalopoulos, "Disaster Avoidance in the Manhattan Fiber Distributed Data Interface Network," Globecom'93, Houston, TX, pp. 680–685, December 2, 1993.
2. S.V. Kartalopoulos, "Surviving a Disaster," *IEEE Communications Mag.*, vol. 40, no. 7, pp. 124–126, July, 2002.
3. S.V. Kartalopoulos, "Security of reconfigurable FSO Mesh Networks and Application to Disaster Areas," SPIE Defense and Security Conference, March 16–20, 2008, Orlando, Florida, paper no. 6975–9, Session S2; Proceedings on CD-ROM.
4. S.V. Kartalopoulos, *Next Generation Intelligent Optical Networks*, From Access to Backbone, Springer, 2008.

MESH FSO SYSTEMS

5.1 INTRODUCTION

In the previous chapters, we examined how Free Space Optical technology can be applied to point-to-point and to ring topologies. In this chapter we examine the mesh topology.

The mesh topology is more complex than the ring and the PtP. Therefore, we first examine the motivation for mesh topology.

The PtP topology, although very successful for rapidly establishing a communications link between two points with line of sight (LoS), by definition is limited because, first, it cannot connect more than two points (or nodes), and secondly, the two nodes cannot have an arbitrary distance between them. If the distance between them exceeds the maximum allowable (in terrestrial applications this is typically $\leq 4\,\mathrm{km}$), then intermediate repeater nodes (with or without add-drop) may be used. However, the intermediate nodes add to network complexity, cost, service and network protection strategy, and maintenance, and therefore their applicability needs to be justified. The PtP topology has the poorest network and traffic protection; if the link becomes inoperable, communications between the two nodes is lost.

Free Space Optical Networks for Ultra-Broad Band Services, First Edition. Stamatios V. Kartalopoulos.
© 2011 Institute of Electrical and Electronics Engineers. Published 2011 by John Wiley & Sons, Inc.

The ring topology is capable to connect more than two nodes, although each node on the ring is an add-drop node, and one or two of the nodes on the ring comprise a bridge via which the ring is connected to other networks. The distance between nodes may be short, yet, the periphery of the ring may be long. The ring topology has better network and traffic protection.

The mesh topology is capable to interconnect many nodes and although, depending on application, the distance between nodes or *inter-node links* (INL) may be as short as few hundred meters or as long as few kilometers, the overall network may extend over many kilometers-square. On the negative side, the complexity (design and maintenance) of each node in the mesh topology increases as the number of INLs per node increases, and thus the network cost-efficiency; to optimize the latter, *topology engineering* or *topology control* is required, which based on algorithms identifies the optimum node interconnectivity. On the positive side, the mesh topology has been extensively studied, is applicable to multi-node networks, is scalable and expandable, is capable of transporting ultra-high volume of standardized traffic at very high data rates (for FSO applications up to 10 Gb/s per link), and provides the best network and service protection. Because of the inherent scalability and expandability, and with *dynamic topology engineering*, the mesh-FSO topology is applicable to ad-hoc networks as long as line of sight is maintained.

5.2 FSO NODES FOR MESH TOPOLOGY

Consider the topography with possible nodes as shown in Figure 5.1. Notice that some nodes in an arbitrary mesh topology may have two, three, four or more ISLs.

Figure 5.1. Nodes over a geographical territory with line of sight between them.

Clearly, as the number of ISLs per node, N, increases, so does the complexity of each node. The reason is that no terrain is completely flat and all nodes are not on the same horizontal plane. Therefore, each ISL is in a different direction than another ISL of the same node. Now, since an ISL needs to always maintain LoS, for a multi-ISL node maintaining ISL tracking is more complicated and costlier as the number of ISLs increases. The question, therefore, is: given the number of nodes on a topology with LoS between them, how can one draw the most cost-efficient network? Clearly, what we are looking for is a network that is constructed with nodes having as few as possible ISLs, and it provides full connectivity among all nodes.

Before we try to answer this, first let us examine the parameters pertinent to this problem.

5.2.1   Parameters Pertinent to FSO Mesh Topology

The challenge here is: given the location of nodes and the topology, identify an optimum node connectivity so that all nodes are interconnected by at least two ISLs, all links are full-duplex (bidirectional), the maximum link length between nodes is not violated, and the number of ISLs per node does not exceed a predetermined number; N may be imposed by design complexity and cost-performance requirements; in the specific case, for which a link is longer that the maximum permissible length, a repeater node may be considered.

To simplify the analysis of mesh-FSO topology, a first degree approximation could be made whereby the terrain is considered flat; this is a reasonable assumption if the mesh network is within few square-miles over a reasonably smooth terrain. We also assume that each node is at different height from the ground level and thus, the laser beams are at different levels and direction in a frame of reference, Figure 5.2. This case is more realistic than assuming that all nodes are on the same flat plane, and thus the actual angle formed by a laser beam at some nodes is more acute, the actual link length may be longer, and the pointing and tracking mechanism should have a wider dynamic

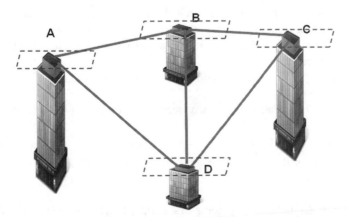

Figure 5.2. Nodes of a mesh-FSO topology on flat terrain but wih nodes at different heights.

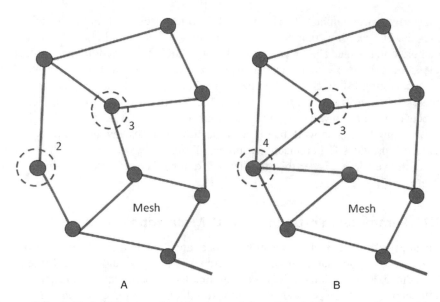

Figure 5.3. A. Mesh-FSO network with two, and three ISLs. B The same network with two, three and four ISLs.

range. Accordingly, the analysis and optimization of realistic mesh-FSO networks becomes a more complex geometrical one.

Thus, given the location of nodes, one needs to interconnect them; clearly, there are several choices: Figure 5.3, illustrates two possible mesh networks for the same nodes: network B consists of nodes with two, three and four ISLs, whereas network A consists of nodes with two and three ISLs.

However, in many cases, the assumption of a smooth terrain is not correct, particularly on uneven and mountainous terrain, and with buildings of different heights. In this case, one needs to first define a three-dimensional Cartesian frame of reference, Figure 5.4 at each node; applying this to Figure 5.2, it becomes evident that each node is displaced with respect to the different frame of reference, angles may be more acute and link lengths longer.

Based on this, now, one may define the following basic parameters:

- *Horizontal distance*: the distance between two nodes projected on the ground plane.
- *Azimuthal angle*: the angle from a horizontal plane that traverses the node.
- *Lateral angle*: the angle from a vertical plane parallel to the plane ZOX that traverses the node.

Having defined these basic parameters, certain basic rules need to be defined that will help construct the optimum mesh network:

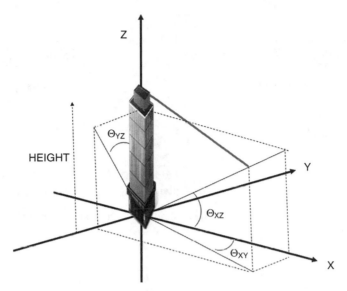

Figure 5.4. Frame of reference in FSO mesh network.

- Maximum horizontal distance between two nodes.
- Minimum azimuthal angle and minimum lateral angle between two ISLs on the same node.
- Maximum number of ISLs per node, N_{max}.

Then, the tolerances may also be incorporated in the last set of parameters that help to determine the quality of service under adverse atmospheric conditions. For example:

- Maximum allowable penalty on horizontal distance due atmospheric effects (fog, rain, etc).
- Maximum allowable penalty on the azimuthal and lateral angles due to wind.
- Maximum allowable penalty on the plane height of the node as a result of post, tower, or building expansion/contraction.

In the above, how ground vibrations may affect the position of each ISL of the node may also be considered.

5.2.2 Mesh-FSO Node Design

Analysis of several mesh-FSO topologies discerned that three, and maybe four, ISLs per node is a satisfactory number, while still maintaining acceptable cost, performance and efficiency of the network. More complex networks that require more ISLs per node

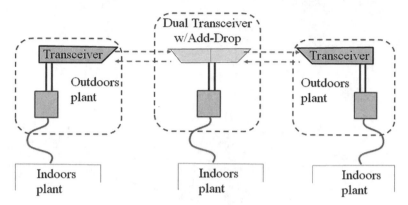

Figure 5.5. Mesh-FSO head constructed with individual PtP transceivers is similar to a node with add-drop features.

increase the complexity and the cost so that the network may become cost-prohibitive; this is minimized with proper topology survey, engineering and control [1–4].

The design of a mesh-FSO node consists of simple PtP transceiver heads that terminate a link whereas the switching or routing function is done at the server or switch of the node, which may not be collocated with the heads (typically located in the indoors plant), Figure 5.5. This allows for flexibility in mounting each head separately, but not the same environmental conditions (temperature, humidity, wind sway, etc.) at each one. However, as integration increases and technology yield smaller parts, two or more transceivers may be collocated within the same protecting housing, assuring the same environmental conditions (temperature, humidity, etc) and requiring a single mounting pole, but it requires the highest point on a building to obtain 360° panoramic view.

Based on this, a mesh-FSO node may be integrated, but the optics and optoelectronic parts (lens, laser, photodetector) that could all be in a very small housing or in individual housings, whereas the electronic switching functionality is in a small assembly that requires a smaller cabinet; smaller component footprint allows for smaller thermoelectric cooling/heating (TEC) devices to better control the environmental condition within housings and cabinets [5]. Figure 5.6 shows a three full-duplex ISL node (but tracking and control mechanisms). Integrated multi-heads are also suitable in ad-hoc networks, as already discussed [6, 7].

5.2.3 Mesh-FSO Network Protection

In a mesh network, a fault may be related to a link (a transmitter, receiver or link amplifier/repeater), to a node (a misbehaving node due to software/control unit misbehaving), or to a cluster of nodes (as in a disaster). The link terminating units have monitoring/detecting mechanisms to identify faults [8], and in some cases to identify severe degradation of traffic performance [9]. Overall, the network is intelligent to identify failing clusters of nodes (a disaster case) and to avoid it [10, 11].

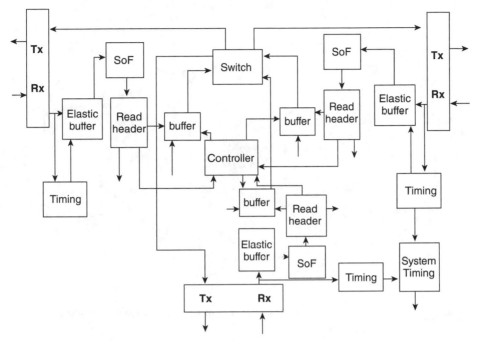

Figure 5.6. A three full-duplex ISL node with key functions.

That is, mesh networks have been thoroughly studied and they have been proven to have superb service, node and network protection. Again, this stems from the network's ability to monitor and detect faults at various levels (link, node, cluster) and automatically reroute traffic to other routes and nodes according to a rerouting algorithm that recalculates the optimum number of hops across the network, maintains traffic balancing so that all nodes in the network deliver traffic at the acceptable performance, and it maintains the traffic priorities, according to service agreements.

5.2.4 Mesh-FSO Scalability

Network scalability refers to the network ability and flexibility to add or delete a node, while service is provided. The mesh topology, in general, is very good in adding/deleting nodes. This stems from its capability to maintain service when single or multiple faults occur, by rerouting traffic over less utilized links and nodes. Thus, when a node is added to the mesh network, the network is configured to recognize the new addition, either by *provisioning* or by *self-discovery* (depending on the protocol that the network employs), and as soon the new node is recognized and is configured to pass traffic, the new node becomes part of the overall mesh network, Figure 5.7.

Conversely, when a node is removed from the mesh network, for all practical purposes it may be considered as a "faulty" node, and the traffic that was previously

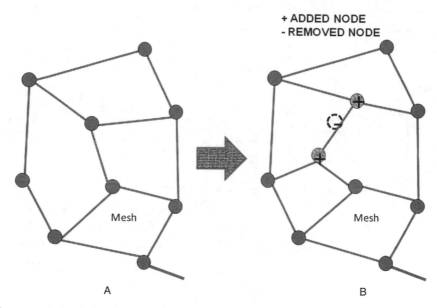

Figure 5.7. **A.** Mesh-FSO network. B. Mesh-FSO network with added and removed nodes.

passing through the node is now rerouted to other nodes while maintain traffic balancing and avoiding traffic congestion; however, in this case, the adjacent nodes to the removed one are provisioned not to issue alarms, and nodes in the network readjust the number of nodes and interconnects in the network, since the network traffic capacity has been reduced by one node and by its interconnecting links.

Mesh-FSO networks, in terms of traffic capacity, traffic routing, and traffic balancing are similarly studied. However, the added variable that makes traffic rerouting and balancing more challenging is that FSOs frequently suffer from atmospheric conditions and thus a cluster of nodes within the mesh network may be degraded as a result of fog set over the cluster of FSO nodes.

An extension to the mesh-FSO is ad-hoc networks. Clearly, nodes may be added and deleted, and moving nodes may change location in the three-dimensional space, thus continuously varying the basic parameters of the network. Therefore, in this case line-of-sight and fast tracking mechanisms make the physical design of the network more challenging.

5.3 HYBRID MESH-FSO WITH RF

As already discussed, FSO technology suffers by atmospheric effects (fog, rain, snow, etc). When fog or other condition degrades the FSO link performance to unacceptable level, then an RF back-up link substitutes the optical link and it maintains the minimum acceptable basic service for the duration of the condition [12, 13].

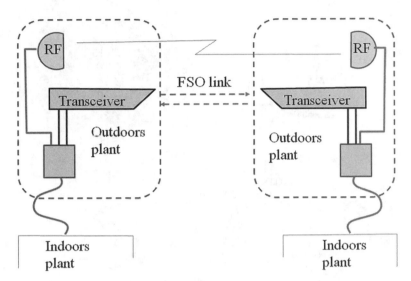

Figure 5.8. FSO with RF back-up.

The mesh-FSO topology is not immune to atmospheric conditions, and therefore it is expected that one or more links in the network will occasionally perform below acceptable level. In this case, RF link back-up may be considered, in which case a traffic back-up algorithm should be used [14]. However, the question is: do we need RF back-up for every link in the mesh network?

The answer is not simple, because it depends on the number of nodes in the network, the number of links per node, the density or the sparsity of nodes in the network, the geographical size of the network, and the fault and disaster avoidance strategy of the network. That is, it may be that all ISLs in the network need back-up, or that few strategic ISLs may be backed-up to construct an overlay of RF links in the network, which in conjunction with a fault avoidance mechanism it may provide service, Figure 5.8.

5.4 HYBRID FSO-FIBER NETWORKS

In some cases, the mesh-FSO network may consist of many nodes. However, if the nodes spread over an urban and a suburban area, then, most likely there is a fiber network within the urban area. In this case, a hybrid FSO-Fiber topology may be considered, so that a simple FSO network acts as one or more extensions to the fiber network that accesses remote nodes or nodes not easily accessible by the fiber network.

Thus, the hybrid FSO-Fiber topology is applicable to cases for which fiber is already in place (most likely a fiber ring topology) and ready to serve some of the nodes of the enterprise network, but not all. In this case, the FSO extension(s) may be as simple as a point-to-point, a small ring or even a small mesh, which may also be

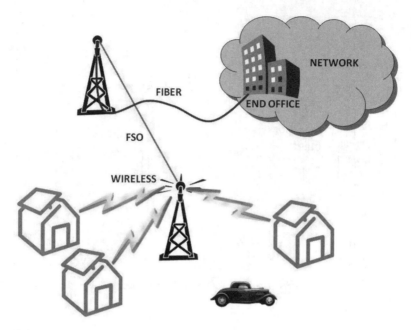

<u>Figure 5.9.</u> FSO as an extension of a fiber network may also be integrated with wireless access to allow mobility to end-users.

integrated with wireless in the access domain to allow for end-user mobility (see Chapter 7), Figure 5.9.

REFERENCES

1. I.F. Akyildiz, X. Wang, and W. Wang, "Wireless mesh networks: a survey," *Computer Networks*, vol. 47(4), pp. 445–487, 2005.

2. P.C. Gurumohan and J. Hui, "Topology design for free space optical networks," *ICCCN*, 2003.

3. A. Kashyap, M. Kalantari, K. Lee, and M. Shayman, "Rollout algorithms for topology control and routing of unsplittable flows in wireless optical backbone networks," Conference on Information Sciences and Systems, 2005.

4. A. Kashyap, S. Khuller, and M. Shayman, "Topology control and routing over wireless optical backbone networks," Conference on Information Sciences and Systems, 2004.

5. S.V. Kartalopoulos, "Free Space Optical Mesh Networks For Broadband Inner-city Communications," NOC 2005, 10th European Conference on Networks and Optical Communications, University College London, July 5–7, 2005, pp. 344–351.

6. A. Desai, J. Llorca, and S. Milner, "Autonomous reconfiguration of backbones in free space optical networks," *IEEE MILCOM*, pp. 1226–1232, 2004.

7. A. Desai and S. Milner, "Autonomous reconfiguration in freespace optical sensor networks," *IEEE JSAC Optical Communications and Networking Series*, vol. 23, no. 8, pp. 1556–1563, 2005.

8. S.V. Kartalopoulos, *DWDM: Networks, Devices and Technology*, IEEE/Wiley, 2003

9. S.V. Kartalopoulos, *Optical Bit Error Rate: An Estimation Methodology*, IEEE/Wiley, 2004.

10. S.V. Kartalopoulos, "Bidirectional Mesh Network," Issued 2/25/1997, 5,606,551.

11. S.V. Kartalopoulos, "Surviving a Disaster," *IEEE Communications Mag.*, vol. 40, no. 7, July 2002, pp. 124 126.

12. H. Izadpanah, T. Elbatt, V. Kukshya, F. Dolezal, and B.K. Ryu, "High-availability free space optical and RF hybrid wireless networks," *IEEE Wireless Networks*, vol. 10, no. 2, pp. 45–53, 2003.

13. A. Kashyap, A. Rawat, and M. Shayman, "Integrated backup topology control and routing of obscured traffic in hybrid RF/FSO networks," *IEEE Globecom*, 2006.

14. A. Kashyap and M. Shayman, "Routing and traffic engineering in hybrid RF/FSO networks," *IEEE International Conference on Communications (ICC)*, 2005.

6

WDM MESH-FSO

6.1 INTRODUCTION

Wavelength division multiplexing (WDM) is an optical technology whereby one or more optical channels, as defined by individual wavelengths, are optically multiplexed and all together are coupled onto a single fiber. Unsurprisingly, this technology first found home in fiber-optic communications [1–4]. Doing so, the amount of data that can be transported over a single fiber is multiplied by the number of optical channels. For example, if 80 optical channels are multiplexed and coupled onto a single fiber, whereby each optical channel transports 10 Gb/s data, then there is an aggregate $80 \times 10 = 800$ Gb/s data rate.

WDM is a technology for which standard documents recommend and define most attributes and specifications, including spectral bands and center wavelengths for each channels. At the end of this chapter we describe two versions of WDM technology, a dense wavelength division multiplexing (DWDM), and a coarse wavelength division multiplexing (CWDM).

Because of the multiplicative nature of the aggregate data rate in WDM technology, and although it was first applied to fiber-optic networks, WDM may also be

Free Space Optical Networks for Ultra-Broad Band Services, First Edition. Stamatios V. Kartalopoulos.
© 2011 Institute of Electrical and Electronics Engineers. Published 2011 by John Wiley & Sons, Inc.

applied to FSO networks. However, in the latter case, one should remember that the fiber is a guided optical medium with parameters that are well-known and are averaged over a long fiber length and over a long period, whereas the atmosphere in FSO is an unguided optical medium with variable parameters over a short length and in a short period.

Although the physics of light-matter interaction is generally known, the interaction of light and atmosphere is a challenge because the molecular consistency, molecular density, temperature, and so on, is continuously changing and therefore it is difficult to model as an optical signal propagating medium. Only on a macroscopic level and for a particular link and time of day some of the medium parameters, such as attenuation, can be measured; this implies that, for the same link and time of day, attenuation may be different for the first 10 meters than the next 10 meters, and so on.

Therefore, in order to be able to evaluate the WDM technology in FSO systems, we briefly look into the WDM technology in fiber-optic systems and compare with FSO.

6.2 LIGHT ATTRIBUTES

The quantum mechanical quantity of light is the photon. Photons are described by Maxwell's theory of electromagnetic wave propagation, on which light interference is based, and also by Plank's theory that photons behave like mass-less particles; that is, light has two natures, wave and particle. In fact, the dual nature of electromagnetic waves can be generalized, although long-wavelength waves behave more like waves and less like particles, ultra-short electromagnetic waves behave less like waves and more like particles (e.g., gamma-rays), and visible light (including infra-red and ultra-violet) behave about the same in both natures. Thus, photons are characterized by a dual nature, and as they propagate in free space or through a medium, their characteristics are affected. Moreover, because light is used in optical communications, the optical channel is not strictly monochromatic (that is, it is not a single wavelength) but consists of many wavelengths in a very narrow band of optical frequencies; only a single photon can be considered a single wavelength. The attributes of light and their significance are listed in Table 6.1:

6.3 OPTICAL MEDIA

When light enters matter, its electromagnetic field interacts with localized electromagnetic fields. The result is a change in the characteristics of light and in certain cases a change in the properties of matter as well. The strength of the field of light, as well as its wavelength, polarization state, and matter characteristics (dielectric constant, density, etc) determine how the light propagation is affected. Additionally, external temperature, pressure and fields (electrical, magnetic, and gravitational) may influence the interaction of light with matter.

TABLE 6.1. Attributes of light in communications

Attribute	Significance
dual nature:	electromagnetic wave and particle ($E = h\nu = pc^2$)
polarization:	circular, elliptic, linear (TE_{nm}, TM_{nm}); it is affected by fields & by matter (polarization change, polarization dispersion)
optical power:	wide range (from μW to MW); it is affected by the dielectric medium
propagation:	in free space in straight path; in matter it is affected (absorbed, scattered, polarization changes, speed changes, possible chromatic dispersion, possible four-wave mixing); in optical waveguides (fiber) it follows its bends
Optical channel consists of many λs:	continuous spectrum; possible chromatic dispersion effects; possible four-wave mixing effects
phase:	affected by variations in fields & matter

6.3.1 Homogeneous and Heterogeneous Media

A *homogeneous* optically transparent medium has the same consistency, chemical, mechanical, electrical, magnetic, and crystallographic throughout its volume and in all directions.

A *heterogeneous* optically transparent medium does not have the same consistency (chemical, mechanical, electrical, magnetic, or crystallographic) throughout its volume.

Based on this, in general, the atmosphere cannot be considered a homogeneous medium over a long distance (hundreds of meters).

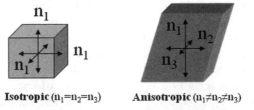

Isotropic ($n_1 = n_2 = n_3$)   Anisotropic ($n_1 \neq n_2 \neq n_3$)

Figure 6.1. Isotropic and anisotropic dielectric media.

6.3.2 Isotropy and Anisotropy

Isotropic optically transparent materials have the same index of refraction, same polarization state and same propagation constant in every direction throughout the material. Materials that do not exhibit all these properties are known as *anisotropic*, Figure 6.1.

Anisotropy is explained as follows: the electrons of certain crystals, such as calcite-$CaCO_3$, move with different amount of freedom in selective directions in the crystal, and thus, the dielectric constant as well as the refractive index of the crystal is different in these selective directions. As a result, as photons enter the crystal their electromagnetic field interacts differently in one direction than in another, and this affects their propagation through the crystal.

In FSO, light propagates through the atmosphere which, over the distance travelled, cannot be considered isotropic.

6.3.3 Propagation of Light in Transparent Dielectric Medium

As already described, when light enters matter, it is reflected by its surface, is refracted by the matter, and its velocity changes as well as its wavelength, but not its frequency; this is important to remember throughout this book.

6.3.3.1 Phase Velocity A monochromatic (single ω or λ) wave that travels along the fiber axis is described by:

$$E(t, x) = A \exp[j(\omega t - \beta x)]   \qquad 6.1$$

Where A is the amplitude of the field, $\omega = 2\pi f$, and β is the propagation constant.

Phase velocity, v_ϕ, is defined as the velocity of an observer that maintains constant phase with the traveling field, i.e. $\omega t - \beta x$ = constant.

Replacing the traveled distance x within time t, $x = v_\phi t$, then the phase velocity of the monochromatic light in the medium is:

$$v_\phi = \omega/\beta   \qquad 6.2$$

6.3.3.2 Group Velocity When a signal is transmitted in a medium, it is necessary to know its speed of propagation. A continuous sine wave does not provide any meaningful information because a real optical signal consists of a band of frequencies in a narrow spectrum. Moreover, each frequency component in the band travels (in the medium) with slightly different phase. Thus, each component travels in the medium at slightly different phase velocity ($\beta_c \pm \Delta\beta$) accruing a different phase shift (where β is the propagation constant).

Group velocity, $v_g = c/n_g$, is defined as the velocity of an observer that maintains constant phase with the group envelope of the propagating signal in the medium. The group velocity is expressed by:

$$v_g = \omega/\Delta\beta = 1/\beta'   \qquad 6.3$$

where β' is the first partial derivative with respect to ω.

6.4 INTERACTION OF LIGHT WITH MATTER

When a photon meets matter, its electromagnetic field interacts with the atoms and molecules of matter, and depending on consistency and structural details of matter,

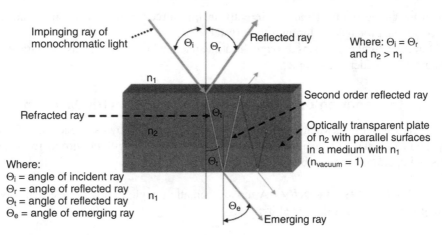

Figure 6.2. Definition of reflection and refraction of optical monochromatic rays.

these interactions may affect the photon properties and the material properties. In many cases, light affects matter, which in turn affects light.

6.4.1 Reflection and Refraction—Snell's Law

When a photonic ray impinges the interface of two homogeneous transparent media (e.g. free space and glass, two layers of different temperature) then a portion of it will be reflected and the remainder will be refracted, Figure 6.2.

Index of refraction of a transparent medium (n_{med}) is defined as the ratio of the speed of monochromatic light in free space, c, over the speed of the same monochromatic light in the medium (v_{med}).

$$n_{med} = c/v_{med} \qquad\qquad 6.4$$

Then, between two mediums (1 and 2) with indices n_1 and n_2, and speed of propagation in corresponding media v_1 and v_2, the following relationships hold:

$$n_2/n_1 = v_1/v_2 \qquad\qquad 6.5$$

and

$$n_1 \cos\beta = n_2 \cos\alpha \qquad\qquad 6.6$$

Where, in a general case, α and β are the angle of incidence and the angle of refraction respectively.

When the angle of incidence is very small, $\cos\alpha = 1 - \alpha^2/2$ and the cosine equation is simplified to:

$$n_1(1-\beta^2/2) = n_2(1-\alpha^2/2) \qquad\qquad 6.7$$

The index of refraction, or *refractive index*, for free space has the numerical value of 1, whereas for other materials it is typically between 1 and 2, and in some cases greater than 2 or 3. (The index of refraction may be negative in a class of materials known as "*metamaterials*").

The reflected portion of monochromatic light is known as Fresnel reflection. The amount of reflected power as well as the polarization state of the reflected light depends on the polarization state of the incident light, on the angle of incidence and on the refractive index difference. For normal incidence on a single surface the reflectivity, ρ, is given by the Fresnel equation:

$$\rho = (n-1)^2 / (n+1)^2 \qquad\qquad 6.8$$

If the absorption of the material over a length d is A, as calculated from the absorption coefficient (absorbed power per km) α, then the internal material transmittance, τ_i, is defined as the inverse of the material absorption.

The following basic relationships are also useful:

speed of light in free space: $c = \lambda f$
wavelength in a medium $\lambda_{med} = c/[f\sqrt{\varepsilon}]$
speed of light in medium: $v_{med} = \lambda_{med} f$
index of refraction: $n_1/n_2 = \lambda_2/\lambda_1$

where f is the frequency of light and e is the dielectric constant of the medium. Usually, either letter f or v are used for frequency. Here, we use f to avoid confusion between v (for speed) and v (for frequency).

6.4.2 Polarization of Light and Matter

The polarization state of propagating photons and the dielectric matter interact in a way that affects the propagation properties of light.

6.4.2.1 Polarization Vector
The electrical state of matter on a microscopic level consists of charges, the distribution of which depends on the presence or not of external fields. Assuming that for every positive charge there is a negative, then each positive-negative charge constitutes an electric dipole. The electric moment of a dipole at some distance is a function of distance and charge density. Now, for a distribution of electric dipoles, *the electric dipole moment per unit volume is defined as the polarization vector* P.

6.4.2.2 Transverse Wave
Two planar relations describe the propagation of light in non-conducting media:

$$E(r,t) = \epsilon_1 E_o \ e^{-j(\omega t - k \cdot r)}, \text{and} \qquad\qquad 6.9$$

$$H(r,t) = \epsilon_2 H_o \ e^{-j(\omega t - k \cdot r)} \qquad\qquad 6.10$$

where ϵ_1 and ϵ_2 are two constant unit vectors that define the direction of each field, k is the unit vector in the direction of propagation, and E_o and H_o are complex amplitudes, which are constant in space and time.

Assuming a wave propagating in a medium without charges, then $(del)E = 0$ and $(del)H = 0$. Based on this, the product of unit vectors is:

$$\epsilon_1 \cdot k = 0 \quad \text{and} \quad \epsilon_2 \cdot k = 0 \qquad\qquad 6.11$$

That is, the electric (E) and the magnetic (H) fields are perpendicular to the direction of propagation k, figure 1a. This wave is called a *transverse wave*.

6.4.2.3 *Circular, Elliptical and Linear Polarization*   Polarization of electromagnetic waves is a complex subject, particularly when light propagates in a medium with different refractive index in different directions, that is, an inhomogeneous medium.

As light propagates through a medium, it enters the fields of nearby dipoles and field interaction takes place. This interaction may affect the strength of the electric and/ or magnetic fields of light differently in certain directions so that the end-result may be a complex field with an elliptical or a linear field distribution.

For example, the electric field E becomes the linear combination of two complex fields E_{ox} and E_{oy}, the two componets in the x and y directions of a Cartesian coordinate system, such that,

$$E(r,t) = (\epsilon_x \ E_{ox} + \epsilon_y \ E_{oy})e^{-j(\omega t - k \cdot r)} \qquad\qquad 6.12$$

This relationship implies that the two components, E_{ox} and E_{oy}, vary sinusoidally, are perpendicular to each other, and that there may be a phase betwen them, ϕ.

Now, from

$$(del)^2 E = (1/\upsilon^2)(\theta^2 E/\theta t^2), \qquad\qquad 6.13$$

and

$$E(r,t) = \epsilon E_o \ e - j^{(\omega t - k \cdot r)} \qquad\qquad 6.14$$

one obtains

$$k \times (k \times E_o) + \mu_o \epsilon \omega^2 E_o = 0 \qquad\qquad 6.15$$

or:

$$[k \times (k \times I) + \mu_o \epsilon \omega^2][E_o] = 0 \qquad\qquad 6.16$$

where I is the identity matrix. The latter is a vector equation equivalent to a set of three homogeneous linear equations with unknowns components of E_o, E_{ox}, E_{oy} and E_{oz}; in the typical case, the component E_{oz} along the axis of propagation is equal to zero. This equation determines a relationship between the vector k (k_x, k_y, k_z), the angular frequency ω, and the dielectric constant ε (ε_x, ε_y, ε_z), as well as the polarization state of the plane wave.

Now, the term $[k \times (k \times I) + \mu_o\varepsilon\omega^2]$ describes a three dimensional surface. As the (complex) electric field is separated into its constituent components, each component may propagate in the medium at a different phase. The phase relationship as well as the magnitude of each vector defines the *mode of polarization*, as follows:

- If E_{ox} and E_{oy} have the same magnitude and are in phase, then the wave is called *linearly polarized*.
- If E_{ox} and E_{oy} have a phase difference (other than 90°), then the wave is called *elliptically polarized*.
- If E_{ox} and E_{oy} have the same magnitude, but differ in phase by 90°, then the wave is called *circularly polarized*.

For example: in circularly polarized light, the wave equation (propagating in the z direction) becomes:

$$E(r,t) = E_o(\varepsilon_x +/- j\varepsilon_y)e^{-j(\omega t - k \cdot r)} \qquad 6.17$$

Then, the two real components (in the x and in the y directions) are:

$$E_x(r,t) = E_o \cos(k \cdot r - \omega t), \text{ and} \qquad 6.18$$

$$E_y(r,t) = -/+ E_o \cos(k \cdot r - \omega t) \qquad 6.19$$

These equations indicate that at a fixed point in space the fields are such that the electric vector is constant in magnitude but it rotates in a circular motion at a frequency ω. Additionally:

- The term $\varepsilon_x + j\varepsilon_y$ indicates a counter-clockwise rotation (when facing the oncoming wave), and this wave is called *left circularly polarized* or a wave with *positive helicity*.
- The term $\varepsilon_x - j\varepsilon_y$ indicates a clockwise rotation (when facing the oncoming wave), and this wave is called *right circularly polarized* or a wave with *negative helicity*.

Using the notion of positive and negative helicity, then, E can be re-written as:

$$E(r,t) = (\varepsilon_+ E_+ + \varepsilon_- E_-)e^{-j(\omega t - k \cdot r)} \qquad 6.20$$

where E_+ and E_- are complex amplitudes denoting the direction of rotation. Now,

- if E_+ and E_- are in-phase but they have different amplitude, the last relationship represents an *elliptically polarized* wave with principal axes of the ellipse in the directions ε_x and ε_y. Then, the ratio semi-major-to-semi-minor axis is $(1 + r)/(1 - r)$, where $E_-/E_+ = r$.
- If the amplitudes E_+ and E_- have a difference between them, $E_-/E_+ = re^{j\alpha}$, then the ellipse traced out by the vector E has its axes rotated by an angle $\phi/2$.
- If $E_-/E_+ = r = +/-1$, then the wave is *linearly polarized*.

The above discussion for the electric field E can also be repeated for the magnetic field H.

In simple terms: The electromagnetic wave nature of monochromatic light implies that the electric and the magnetic field are in quadrature and in time phase. When created light propagates in free space, the two fields change sinusoidally and each one lies on one of two perpendicularly to each other planes, as in Figure 6.3.

When light enters matter, then depending on the displacement vector distribution in matter (and hence the dielectric and the refractive index), the electric and/or magnetic field of light interacts with it in different ways. If the planes of the electric or the magnetic field are in a Cartesian coordinate system and fixed, then light is *linearly polarized*. If on the other hand, the planes keep changing in a circular (corkscrew-like) motion and the (vectorial) fields remain at the same intensity, then light is *circularly polarized*. If in addition to it, the intensity of the field changes monotonically, then light is *elliptically polarized*. Figure 6.4 illustrates some polarization distributions around the axis of propagation of an electromagnetic plane wave.

Now, if light is separated into two components, one linearly polarized, I_P, and one unpolarized, I_u, then, the degree of polarization, P, is defined as:

$$P = I_P/(I_P + I_U) \qquad 6.21$$

Light may be polarized when it is reflected, refracted, or scattered.

In polarization by reflection, the degree of polarization depends on the angle of incidence and on the refractive index of the material, given by the *Brewster's Law*:

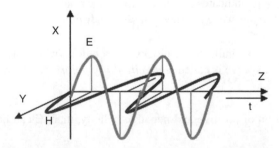

Figure 6.3. Propagation of the E-M wave for an observer at the origin.

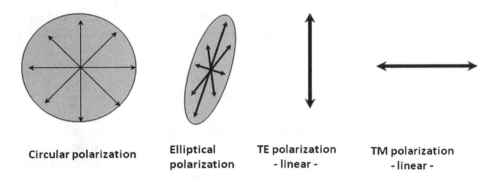

Circular polarization   **Elliptical polarization**   **TE polarization - linear -**   **TM polarization - linear -**

TE=Trans-Electric
TM=Trans-Magnetic

Figure 6.4. Circular, elliptical and linear polarizations.

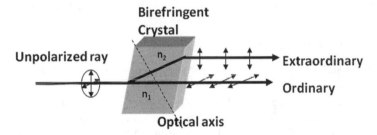

Figure 6.5. Birefringent materials split the incident beam in two rays, each with different polarization.

$$\tan(I_P) = n \qquad\qquad 6.22$$

where n is the refractive index and I_P the polarizing angle.

6.5 MEDIUM BIREFRINGENCE

Anisotropic media have a different index of refraction in specific directions. As such, when a beam of monochromatic unpolarized light enters the material in a specific angle and travels through it, it is refracted differently in the directions of different indices, Figure 6.5. That is, when an unpolarized ray enters the material, it is separated into two rays, each with different polarization, different direction, and different propagation constant; one is called *ordinary* ray (O) and the other *extraordinary* ray (E). In these directions, the refracted index is similarly called *ordinary index*, n_o, and *extraordinary index*, n_e, respectively. This property of optical media is known as *birefringence*.

In general, all optically transparent media have some degree of birefringence; crystals have much higher birefringence than the atmosphere because of the much higher molecular density [5].

6.6 DWDM AND CWDM OPTICAL CHANNELS

FSO technology has used the same optical frequencies used in fiber-optic communication, in the wavelength spectrum 1260–1625 nm taking advantage of existing telecom components, and also the additional spectrum 780–880 nm taking advantage of inexpensive *vertical cavity surface emitting laser* (VCSEL) devices.

The fiber-optic spectrum 1280–1620 nm and the optical channel in it has been defined by ITU specifications for two different optical telecommunication applications, the dense wavelength division multiplexing (DWDM) and the coarse wavelength division multiplexing (CWDM). DWDM is more precise and thus more expensive than CWDM and it is better suited in long-haul (many km) communications. CWDM is less expensive and it is better suited in short-haul (several km) communications. As such, CWDM technology provides a better match, considering distance and cost, to FSO.

6.6.1 The DWDM Grid

The spectral band 1260–1625 nm for optical communications has been partitioned in five ranges.

The wavelength range 1528–1561 mn is known as the *C-band* and the range 1561–1660 nm as the *L-band*, and these two are currently much used in fiber-optic networks. The band 1260–1360 nm (known as the O-band) has not been used yet (at the exception of its wavelength 1310 nm), nor the bands 1360–1460 nm (known as the E-band) and 1460–1528 nm (known as the S-Band). These bands are planned for the future and for special applications.

ITU-T recommendation G.692 [6, 7] recommended optical channels (with a center wavelength) that starts from the reference center frequency 196.10 THz (or 1528.77 nm) and then by incrementing or decrementing in multiples of 50 Ghz (or 0.39 nm) the center frequencies of other channels are defined. Doing so, a dense grid of optical channels is defined.

Based on the same reference frequency (196.10 THz or 1528.77 nm), ITU also defines a sparser grid with fewer channels, by incrementing or decrementing by 100 Ghz, 200 Ghz, or 400 Ghz, and also a denser grid by incrementing or decrementing by 25 Ghz. Such a dense grid over the complete optical spectrum yields more than 2000 optical channels.

6.6.2 The CWDM Grid

ITU-T Recommendation G.694.2 has also defined a coarse grid of 18 channels with 20 nm spacing over the spectrum 1261 to 1621 nm. This spacing tolerates large spectral drifts due to device temperature variation and thus it is suitable for cost-effective optical access networks, such as the *Fiber to the Home* (FTTH), and fiber-based small and medium local area networks.

The center frequencies of the CWDM channels are not defined as in the DWDM case but they are set at the wavelengths: **1271, 1291, 1311, 1331, 1351, 1371, 1391, 1411, 1431, 1451, 1471, 1491, 1511, 1531, 1551, 1571, 1591, 1611.**

It is obvious that a subset of these wavelengths can also be applied to FSO WDM networks, point-to-point, ring and mesh. In this case, one or more optical channels may be transparently pass through a node and be terminated by their destination node.

6.7 WDM FSO LINKS

Based on the optical channel grid defined by ITU standards, the CWDM channels may be easily used in FSO links, and particularly the wavelengths 1311 nm, 1531 nm, 1551 nm and 1571 nm.

In FSO, multiplexing optical channels comes down to transmitting two or more optical beams, each at different wavelengths, parallel and through the atmosphere. That is, a multiple channel beam is transmitted towards the same receiver target.

At the receiver, the multiple channel beam is separated into its component channels by a filtering process; this process may incorporate any of the well-known demultiplexing methods.

Multiplexing a number of channels, as in 2^n or (2, 4, or 8), in the FSO link may reduce the actual data rate per channel yet maintaining an effective high data rate. For example, if four channels are used, each channel at an actual 250 Mb/s data rate, then the effective data rate in the beam is 1 Gb/s. Because of the lower data rate per channel, this method [8] is suitable for longer links and/or high performance links.

In addition, WDM networks (ring and mesh) facilitate routing of information, and provide better service availability and survivability [9–13].

6.8 WDM MESH FSO NETWORKS

In this section, we consider the case where a multimode FSO network needs to be established and describe the major activities the engineer needs to take. That is, the communications engineer starts with a "white page" and he/she is called to fill it with a meaningful drawing in detail [14].

6.8.1 Mesh Network Engineering

The challenge presented to the network communications engineer is to study all known parameters related to traffic needs (data rate, network capacity, traffic patterns, number of nodes, etc) and identify all unknown parameters related with topology so that the resulting network meets and exceeds all communication needs, is cost-efficient, offers excellent service protection, and the interconnected nodes have as few as possible links. To accomplish this, one can follow an optimization methodology by which all nodes are interconnected and in the most cost-efficient manner. In addition, a few optical channels (2 or 4) per link may be considered to be used either for lowering the data rate per optical channel, or, to create an overlay of multiple networks; such networks may be viewed as coexisting networks having common nodes but each network using only one of the optical channels.

In the typical case, the number of nodes in the FSO network is small, 4–20, and thus the network can be easily engineered without complex optimization algorithms but with an experiential optimization methodology. FSO networks with many nodes (hundreds) are atypical and in this case, complex optimization algorithm development is but an interesting exercise.

When the parameters of the communication needs have been set, then a network engineering analysis consists of the following major efforts:

- Survey of the area where a possible FSO network is to be deployed.
- Localization of nodes over the area.
- Measure altitude of each node, distances from a node to its adjacent nodes (typically those nodes that are within a maximum distance), and the angles (horizontal and vertical) from a node to its adjacent nodes.
- Locate problem spots in the area (obstructions that hinter line of sight, or angles that exceed the maximum permissible by the positioning mechanism).
- Identification of FSO nodes that will support connectivity with the public network (synchronous or asynchronous).
- Estimation of peak and average data traffic per node and the maximum aggregate traffic over the mesh network.
- Determine (based on traffic volume) whether a single optical channel per node or multiple (WDM) channels will be used.

6.8.2 Link Alignment

A mesh node may have collocated multiple transceivers or transceivers that are physically separated.

Collocated transceivers means that two or more transceivers are located in the same housing. In such case, the transceivers are either in a pile or stacked up (one on top of the other) configuration, or they are coplanar (on the same plane). In the stacked up configuration, each transceiver must have the ability to rotate laterally on its plane within 360 degrees, as well as vertically (up or down) and within an angle $\pm\Theta°$, where Θ is an angle determined by the manufacturer; typically, it should be greater than 45°. In the coplanar configuration, each transceiver must have the ability to rotate laterally on its plane within a sector of an angle of $2\pi/n$, where n is the number of transceivers in the head, as well as vertically (up or down) and within an angle $\pm\Theta°$, where Θ is an angle determined by the manufacturer; typically, it should be greater than 45°.

Separated transceivers means that although the transceivers are on the same rooftop, each transceiver is in its own housing, as if it is a single transceiver.

The link pointing, alignment and tracking mechanisms and procedures remain the same as already described.

6.8.3 Mesh-Network Optimization

Mesh network optimization is an effort that is performed at the initial phases of network engineering topology analysis. For example, assume that a number of possible nodes are identified over a specific terrain, Figure 6.6.

In this case, the simplest network to construct is a ring network, Figure 6.7. However, because the ring network does not exhibit the same service protection level of the mesh network, then the same nodes may be interconnected to construct a mesh network, Figure 6.8. In the latter case, we also made an attempt not to exceed the number of links per node beyond three. We have found that for medium mesh networks,

Figure 6.6. Areal photograph with possible FSO nodes mapped on it.

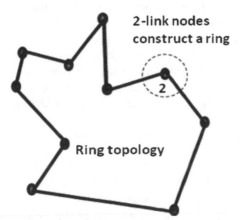

Figure 6.7. The nodes in Figure 6.6 may be connected to form a ring-FSO topology.

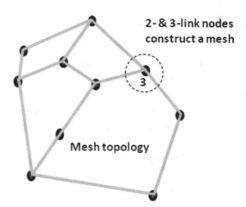

2- & 3-link nodes
construct a mesh

3

Mesh topology

<u>Figure 6.8.</u> The nodes in Figure 6.6 may be connected to form a mesh-FSO topology.

nodes with two and three links are sufficient to construct a cost-efficient mesh FSO network [15].

6.9   SERVICE PROTECTION IN MESH-FSO NETWORKS

Mesh-FSO networks may consider half-duplex or full-duplex links. Full-duplex links have a larger traffic capacity and offer superb protection against node or link failure [16–18]. Half-duplex links simplify node complexity and offer adequate protection and acceptable service. In either case, when a faulty node is detected, traffic is rerouted over a separate path according to well-known rerouting practices and routing algorithms.

In general, mesh-FSO networks employ fault and performance monitoring mechanisms, fault management, as well as protection strategies similar to landline networks. Such mechanisms may be based on performance (bit error rate (BER) and on error-detection, error correction-codes (EDC), or on more advanced estimation methods [19].

When a link fault or severe degradation is detected, the link and node needs to be tested. This is accomplished by performing data loop-backs. Data loop-back is accomplished by provisioning the switching fabric; this is possible with maintenance and controlled packets conforming to the actual protocol in use (e.g., SONET/SDH or GbE).

The following fault/degradation cases may be considered:

- Link degradation due to fog, smoke, misalignment and other causes that degrade link performance below the nominal value for a given bit rate, yet above a minimum acceptable bit rate. This is a situation that yields excessive BER and its degradation trend needs to be tracked; if it is improving (e.g., fog is moving out) or if it worsens (e.g., fog is moving in).
- Link degradation due to fog, smoke, misalignment and so on, causes levels to go below the minimum acceptable level. The degradation is considered severe

and is addressed as link failure although it is tagged accordingly to avoid confusion with permanent link failure.

- Link hard failure due to a faulty link. The failure may be persistent (hard failure, permanent misalignment).
- Degradation of a cluster of nodes caused by atmospheric effects. For example, as fog spreads over a cluster of nodes, their performance is degraded below the expected value for a given bit rate, yet above a minimum acceptable bit rate. Degradation may be within the acceptable level or it may be severe.

6.9.1   Link Degradation

The link performance may be degraded by a number of mechanisms. Wind causes misalignment and thus link degradation. Similarly, airborne particles (fog, snow, smoke, sand, heavy rain) cause signal attenuation.

When link performance is degraded below the expected value for the maximum sustained bit rate, yet above a minimum that is set by network engineering rules, then the node may execute one of two scenarios:

- The node may gradually lower the bit rate in predefined steps, and it may also activate its RF back up link to pass excess traffic.
- The node may switch operation to RF back up link only.

Either strategy requires intelligence and link congestion protocols by which neighboring nodes make appropriate adjustments to balance and/or groom traffic on the network. Additionally, when traffic is split over the degraded FSO link and over the RF back-up link, traffic partitioning may be autonomously determined based on degree of degradation and payload type; one may consider multi-play, by which the link carries all types of payload (voice, data, music, video).

6.9.2   Link Hard Failures

Link hard failure is a permanent condition and until the link is serviced, it is handled as any link failure in mesh networks; that is, traffic is rerouted over other paths, bypassing the failed link. If RF back up is provided, then the RF link may continue with degraded service. Clearly, this case requires traffic balancing and in certain cases low priority traffic may be denied.

Severe degradations are handled like failures, although in this case the expectation is that service will eventually be self-restored as soon as the condition disappears (such as fog or smoke).

6.9.3   Degradation of a Cluster of Nodes

This case affects the throughput of the overall network. If RF link back up is supported, then traffic flow may be separated and part of it may be transported over RF links and the other part over the degraded optical links while maintaining traffic balancing over the network; traffic split can be autonomously determined based on the degree of link degradation or by remote node provisioning.

When severe degradation occurs, then a disaster avoidance protocol is adopted to either bypass the affected network area and/or switch to RF back-up. If there is no RF back-up, then the network may either become partially or totally inoperable.

6.10   WDM MESH-FSO VERSUS EM-WIRELESS

EM-wireless (wireless networks in the electromagnetic spectrum below the VHF) has the unique advantage of end-terminal mobility with or without line of sight (LoS). However, mesh-FSO with LoS has the advantages of bandwidth, link length, and security which in most applications EM-wireless does not have. In some applications, bandwidth is important because it can carry real-time high resolution data, and additionally security is also critical [20].

Although EM-wireless does not deliver the bandwidth the WDM mesh-FSO does, EM-wireless may support the FSO network because it is not as vulnerable to fog and snow as FSO is, as already discussed.

In the following, we make a comparison between the two technologies, Table 6.2.

In conclusion, EM-wireless technology supports mobility better than any other technology. However, this mobility is constrained by limited bandwidth so that it is applicable to compressed voice and other low data-rate applications; higher data rates are possible but the antenna-to-antenna distance is short and they can serve a relatively limited number of devices.

TABLE 6.2. Comparing EM-wireless with WDM mesh-FSO

	EM-wireless	WDM Mesh-FSO	Importance to communications
End user mobility (in general)	high	low	high
Continuous data rate	low	ultra high	high
Packetized data rate	moderate	ultra high	high
(data rate) × (distance)	low	very high	high
security features	low	very high	very high
impairments	some	some; almost none with RF back up	high
ease of deployment	easy	easy	high
network reliability	moderate	high	high
network availability	moderate	high	very high
network robustness	moderate	high	very high
Cost/BW	low	low	Depends on application

In contrast, the WDM mesh-FSO technology supports mobility with LoS, and it supports ultra-high bandwidth that exceeds a continuous 10 Mb/s for many end-devices and for longer distances, while at the same time communication is inherently more secure and more robust. Moreover, if FSO nodes are equipped with RF back up, then WDM mesh FSO networks rival EM-wireless on all merits.

REFERENCES

1. S.V. Kartalopoulos, *Introduction to DWDM Technology: Data in a Rainbow*, IEEE/Wiley, 2000.
2. S.V. Kartalopoulos, *DWDM: Networks, Devices and Technology*, IEEE/Wiley, 2003.
3. S.V. Kartalopoulos, "Ultra-Fast Self-Restoring Optical WDM Channels with Enhanced Service Availability", Proceedings of the 12th WSEAS International Conference on Communications, ISBN: 978-960-6766-84-8, July 2008, pp. 58–61.
4. S.V. Kartalopoulos, "Next Generation Hierarchical CWDM/TDM-PON network with Scalable Bandwidth Deliverability to the Premises", *Optical Systems and Networks*, vol. 2, pp. 164–175, 2005.
5. R.M.A. Azzam and N.M. Bashara, *"Ellipsometry and Polarized Light"*, North Holland, Amsterdam, 1977.
6. ITU-T Recommendation G.652, *Characteristics of a single-mode optical fiber cable*, October 1998.
7. ITU-T Recommendation G.671, *Transmission characteristics of passive optical components*, November 1996.
8. S.V. Kartalopoulos, "Bandwidth Elasticity with DWDM Parallel Wavelength-bus in Optical Networks", *SPIE Optical Engineering*, vol. 43, no. 5, pp. 1092–1100, May 2004.
9. D. Banerjee and B. Mukherjee, "Wavelength-routed optical networks: Linear formulation, resource budgeting tradeoffs, and a reconfiguration study," *IEEE/ACM Transactions on Networking*, vol. 8, no. 5, pp. 598–607, 2000.
10. R.M. Ramaswami and K.N. Sivarajan, "Design of topologies: A linear formulation for wavelength routed optical networks with no wavelength changers," *IEEE/ACM Transactions on Networking*, vol. 9, no. 2, pp. 186–198, 2001.
11. E. Leonardi, M. Mellia, and M.A. Marsan, "Algorithms for the logical topology design in WDM all-optical networks," *Optical Networks Magazine, Premiere Issue*, vol. 1, no. 1, pp. 35–46, 2000.
12. Z. Zhang and A. Acampora, "Heuristic wavelength assignment algorithm for multihop wdm networks with wavelength routing and wavelength re-use," *IEEE/ACM Transactions on Networking*, vol. 3, no. 3, pp. 281–288, 1995.
13. S.V. Kartalopoulos, *DWDM: Networks, Devices and Technology*, IEEE/Wiley, 2003.
14. S.V. Kartalopoulos, "Free Space Optical Nodes Applicable to Simultaneous Ring & Mesh Networks", Proceedings of the SPIE European Symposium on Optics & Photonics in Security & Defense, Stockholm, Sweden, 9/11–16/2006, paper no. 6399-2.
15. S.V. Kartalopoulos, "Protection Strategies and Fault Avoidance in Free Space Optical Mesh Networks", IEEE ICCSC'08 Conference, Shanghai, May 26–28, 2008; Proceedings on CD-ROM: ISBN 978-1-424-1708-7.

16. S.V. Kartalopoulos, "Free Space Optical Nodes Applicable to Simultaneous Ring & Mesh Networks", Proceedings of the SPIE European Symposium on Optics & Photonics in Security & Defense, Stockholm, Sweden, 9/11–16/2006, paper no. 6399-2.

17. S.V. Kartalopoulos, "Surviving a Disaster", *IEEE Communications Mag.*, vol. 40, no. 7, pp. 124–126, July 2002.

18. S.V. Kartalopoulos, "Disaster Avoidance in the Manhattan Fiber Distributed Data Interface Network", Globecom'93, Houston, TX, pp. 680–685, December 2, 1993.

19. S.V. Kartalopoulos, "Circuit for Statistical Estimation of BER and SNR in Telecommunications", Proceedings of IEEE ISCAS 2006, May 21–24, 2006, Kos, Greece; on CD-ROM, paper #A4L-K.2, ISBN: 0-7803-9390-2, Library of Congress: 80-646530.

20. S.V. Kartalopoulos, *Next Generation Intelligent Optical: Networks: from Access to Backbone*, Springer, 2008.

7

INTEGRATING MESH-FSO WITH THE PUBLIC NETWORK

7.1 INTRODUCTION

Currently, FSO technology has been used to transport traffic compliant to standardized protocols such as the next-generation SONET/SDH, the Ethernet, the ATM, and the Internet protocol TCP/IP). These are also protocols that are supported by the public (fiber) network.

It is desirable to use one of the standardized protocols that is also supported by the network provider; this assures interoperability and avoids translating one protocol to another; that is, it avoids the additional step of encapsulation and adaptation. Nevertheless, this is not always possible and therefore translating one protocol to another is a necessity.

It should be pointed out that, the FSO technology is not protocol limited and it can support any synchronous and asynchronous telecommunications and data protocol (e.g., DS1, DS3, IP, Ethernet, SAN, FC), including private protocols; DS1 and DS3 are the digital Service level 1 and 3, respectively, SAN stands for Storage area networks, and FC stands for Fiber Channel.

Free Space Optical Networks for Ultra-Broad Band Services, First Edition. Stamatios V. Kartalopoulos.
© 2011 Institute of Electrical and Electronics Engineers. Published 2011 by John Wiley & Sons, Inc.

Typical fiber-based networks support the next-generation SONET/SDH over WDM. This is not a surprise since the original SONET/SDH protocol, although developed in the 1980s, is still outperforming other modern protocols. Now, the services offered in the 1980s did not have the service variability of today, and therefore, in order to meet current and near future market demands, the original protocol was updated and new extension protocols were developed.

The next-generation SONET/SDH includes extension protocols to transport traditional traffic (DS1, DS3, ATM) as well as data traffic (Ethernet, Internet, etc) with flexible traffic allocation, intelligent routing schemes, elastic bandwidth, multicast capability, better management strategies, future proofed technology, increased network efficiency and cost competitiveness with the next generation data networks.

The original SONET/SDH protocol was developed to support legacy data rates, from DS0 to DS3 and higher, to meet real-time requirements, to be robust, to support fast service protection (50 msec or less). The SONET/SDH supported network topologies were two- or four-fiber rings, point to point, and interconnected rings, which effectively emulated a mesh topology [1].

The SONET/SDH protocol was based on specific size frame-structures, the smallest frame consisted of bytes arranged in a matrix of 90 columns by 9 rows. The first three columns of this frame (called STS-1 for SONET and STM-0 for SDH) were allocated for (section and line) *transport overhead*, and the remaining 87 columns were allocated, one for path overhead, 84 for user payload and two columns were unused (called "fixed stuff"). The payload could be filled by tributary units (TU in SDH) or virtual tributaries (VT in SONET) of specific sizes that contained end user data; the SONET/SDH structure is examined in a subsequent part in this chapter.

The network topology, switch to protection and data-rate objectives for SONET/SDH were met and surpassed, and the initially used data rates of up to 622 Mb/s (OC-12) were extended up to OC-768 (40 Gb/s).

After the initial success of SONET/SDH, wavelength division multiplexing (WDM) made significant inroads in fiber networks. *SONET/SDH over WDM* was a natural solution but it raised certain issues pertaining to traffic efficiency, service flexibility, service protection and cost, losing its competition with exploding data-centric networks, such as the Ethernet, the Internet, etc. This led to a competitive next generation protocol known as *next generation SDH/SONET* (NG-S), which required other new protocols such as the *generic multi-protocol label switching* (GMPLS), *Link Access Procedure SDH* (LAPS), the *Generic Framing Protocol* (GFP) and the *Link Capacity Adjustment Scheme* (LCAS). With the new protocols, the NG-S now is capable to transport over adaptable routes in a variety of network topologies (ring and mesh) synchronous traffic such as voice, video, and asynchronous data [2].

In this chapter, we describe specific popular protocols, among which are the Ethernet, the Internet, and the next-generation SONET/SDH, including GMPLS, LAPS, GFP, and LCAS. In addition, we describe protocol adaptation methods. That is, how one protocol is mapped into another so that an FSO that uses one standard protocol can be mapped onto the standard protocol of the network provider, if the two are not the same; hence integration of FSO with the overall communications network.

7.2 THE ETHERNET PROTOCOL

The initial Ethernet protocol was developed almost three decades ago for a hierarchical local area data network suitable for a tree topology, high data rate, simplicity, relatively short distance transport, quick installation, easily maintainable, and low cost [3].

The Ethernet was accepted as an industry standard (IEEE 802.3) and its popularity keeps growing because it is a public standard, is simple, and its cost keeps decreasing because of technological advancements.

Because the initial Ethernet was not developed to compete with telephony, error control, network protection, security, real-time data delivery, and quality of service were secondary issues. Nevertheless, new Ethernet protocol versions are more service aggressive and at data rates of 1, 10, 40 and perhaps 100 Gb/s, they include mechanisms to transport voice, fast data, and real-time video.

The initial Ethernet protocol was based on the tree topology. Terminal stations on the Ethernet could initiate a frame transfer. Thus, there was a finite probability that two or more stations could attempt a frame transfer at the same time, in which case a collision could occur. To avoid collisions, all Ethernet stations use a carrier sense multiple-access/collision detection (CSMA/CD) mechanism that resolves collisions with "equal fairness to all stations". This mechanism is based on a predefined collision window and on active listening to traffic on the network. If the carrier is absent for at least twice the collision window, a station can transmit. If two or more stations start transmitting, they are able to detect the collision from "listening to the traffic" and then back off. At a random time thereafter, they start again.

Transmission starts with a preamble signal (a string of alternating zeroes and ones). The purpose of the preamble is to listen and to also stabilize the clock of the receivers. The preamble code is followed by a packet; this consists of the start-of-frame delimiter, the source ID, the destination ID and other information, including length of data field; this is followed by user data, and it is concluded with a frame check sequence.

The CSMA/CD of the Ethernet works as follows: a node on the tree network "listens" to traffic on it. When a node needs to transmit, it ceases the network only when it is "quite"; that is, there is no other node transmitting. This is the CSMA part of the protocol. If two nodes try to cease the network simultaneously, then they "sense" each other's attempt (that is, they detect a collision). Then, they both give way and try again but after a random interval. This is the CD part of the protocol. The frame format of the latter is illustrated in Figure 7.1.

There are several Ethernet variants. Although not all of them are currently used in FSO networks, and some are defined for copper cables, for completeness we list them all:

- Ethernet 10BASE-T & 100BASE-T. They are defined for unshielded twisted pair cable (UTP) at 10 & 100 Mb/s.
- Ethernet 1000BASE-x is defined for UTP and fiber at 1000 Mb/s; x = T for UTP and F for fiber
- 10GbE is defined at 10 Gb/s over fiber
- 40GbE is defined at 40 Gb/s over fiber

SFD = start-of-frame delimiter
DA = destination address
SA = source address
LLC = Logical link control
PAD = Packet assembler-disassembler
FCS = frame check sequence
DSAP = Destination service access point
SSAP = Source service access point
The preample consists of at least 70×AA octets

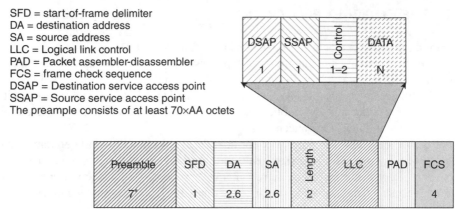

*Numbers indicate octets
0×AA = 10101010

Figure 7.1. Frame format of CSMA/CD Ethernet.

7.2.1   Gigabit Ethernet

The Gigabit Ethernet (1000BASE-x), or GbE, has evolved from the 100 Mb/s Ethernet (100BASE-x) and from the 10 Mb/s Ethernet (10BASE-T). GbE was initially defined for twisted copper cable, known as 1000BASE-T and subsequently for optical fiber [4, 5], it is backward compatible with its predecessor "fast Ethernet" and thus it uses the carrier sense multiple access/ collision detection (CSMA/CD) access method. The GbE is also popular in FSO networks.

Ethernet defines several layers, the Media Access control (MAC), the Physical Medium independent interface (PMI), and the Physical Layer (PHY), which consists of the physical sublayer and the Medium dependent interface sublayer (MDI).

GbE also defines an intermediate layer between the PMI and PHY known as the Medium Independent Interface (MII). The purpose of MII is to provide medium transparency to layers above it and allow for a variety of media (wired, MMF and SMF), as already described.

The MAC layer provides network access controllability during frame assembly and disassembly of the client data and during frame transmission to lower layers. It also provides network access compatibility of MACs, end-to-end and of all intermediate MACs on the path. The MAC layer may optionally provide full or half-duplex capability; full-duplex means that the MAC supports both transmit and receive. The MAC layer may also request that subsequent peer MACs inhibit further frame transmission for a predetermined period, and it may allow for a second logical MAC sublayer. In the latter case, the basic frame format is maintained but the interpretation of the data fields and the length may differ. It may also support Virtual-LAN (VLAN) tagging (per IEEE 802.3ac, 1998) to prioritize packets; however, VLAN tagging requires changes to the frame format.

In the GbE standard, four types of media are defined, 1000Base-SX, 1000Base-LX, 1000Base-CX, and 1000Base-T. The first two (1000Base-SX and 1000Base-LX) con-

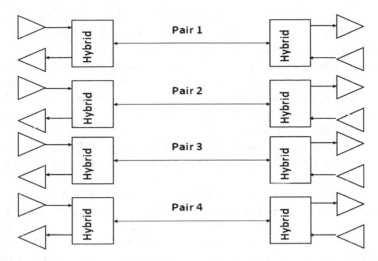

Figure 7.2. Four parallel and synchronized pairs share the total bandwidth.

sider optical fiber as the physical medium and the Fiber Channel (FC) technology for connecting workstations, supercomputers, storage devices and peripherals; the last two (1000Base-CX, and 1000Base-T) consider copper medium.

1000BASE-T is not defined at a serial Gb/s bit rate. Instead, it is defined over four unshielded twisted pairs (UTP), category-5, 100 Ohm copper cables, up to 100 meter maximum length conforming to ANSI/TIA/EIA-568-A cabling requirement, at 250 Mb/s per pair. Conversely, the 1000BASE-T complies with the same topology rules of 100BASE-T and it supports half-duplex and full-duplex CSMA/CD. It also uses the same auto-negotiation protocol with 100BASE-TX. The four pairs form a parallel cable, each keeping the symbol rate at or below 125 Mbaud, Figure 7.2.

Because 1000BASE-T is defined for copper lines, the standard had to address known issues such as echo, near-end crosstalk, far-end crosstalk, noise, attenuation, and EMI. To keep noise, echo and crosstalk at low levels commensurable with 10^{-10} bit error rate, the following countermeasure design strategies were adopted:

1. 4D 8-state trellis forward error correction (FEC) code
2. Signal equalization with digital techniques
3. PAM-5 multilevel encoding where each symbol represents one of five levels, $-2, -1, 0, +1$ and $+2$. The four levels are used for data and the fifth for FEC coding. This results in a reduction of the signal bandwidth by a factor of two.
4. Pulse shaping at the transmitter to match the characteristics of the transmission channel and increase the signal to noise ratio.
5. Scrambling to randomize the sequence of transmitted symbols and reduce spectral lines in the transmitted signal.

Thus, 1000BASE-T may easily be adapted and simplified to four-channel WDM FSO, whereby each channel operates at 250 Mb/s (for better performance, reliability and longer link lengths), whereas the aggregate bit rate is 1 Gb/s. Notice that in this case, items 1 to 4 may be eliminated, or simplified since transmission in FSO is optical.

1000BASE-CX standard defines a Gigabit Ethernet for single-pair "twinax" shielded twisted pair (ST) copper cable but for very short lengths (up to 25 meters). In addition to UTP cable, the 802.3ab task force has defined standards for fiber cable. These standards are the 1000BASE-SX for short-wavelength fiber (850 nm, MMF) and the 1000BASE-LX for long-wavelength fiber (1300 nm, SMF). These standards may also be used in FSO transmission.

1000BASE-LX defines a Gigabit Ethernet for long-haul (up to 3 km) using an optical channel at 1300 nm over single mode fiber, or multi mode fiber (up to 550 meters).

1000BASE-SX defines a Gigabit Ethernet for long-haul (up to 3 km) using an optical channel at 850 nm over multi mode fiber with core diameter 50 μm (up to 550 meters) or with core diameter 62.5 μm (up to 300 meters).

Because of the four physical media defined in GbE (1000BASE-CX, 1000BASE-SX, 1000BASE-LX and 1000BASE-T), it was necessary to bring transparency to the media access control (MAC) layer. Thus, a new layer under the MAC was defined, known as the reconciliation sublayer and the Gigabit media independent interface (GMII), Figure 7.3.

The GMII provides physical layer independence so that the same MAC can be used for any media and it supports 10, 100 and 1000 Mb/s data rates for backward

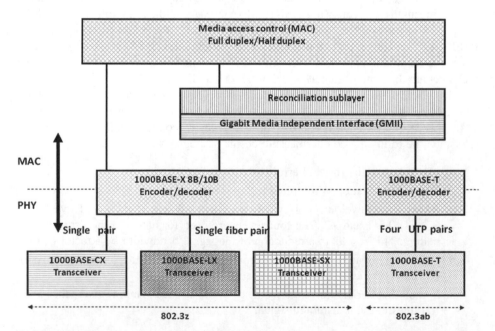

Figure 7.3. The reconciliation sublayer and the Gigabit media independent interface (GMII) between the MAC and the PHY layers.

compatibility. It also includes signals such as management data clock and management data input-output and basic and extended (which is optional) registers. The registers are used for auto-negotiations, power down, loop-back, PHY reset, duplex/half-duplex selection, and others.

The Reconciliation sublayer in the transmit direction maps service primitives or physical layer signaling (PLS) from the MAC to MGII, and vice versa in the receive direction.

The PLS includes signals such as data request, transmit enable, transmit error, transmit clock, collision detect status, data valid, receive error, signal status indicate, carrier status indicate (collision detect, carrier sense, carrier extend), and receive clock.

The GMII is further subdivided into three sublayers, the *physical coding sublayer* (PCS), the *physical medium attachment* (PMA) and *physical medium dependent* (PMD).

- The PCS sublayer provides a uniform interface to the Reconciliation layer for all physical media; it uses the 8B/10B coding like the Fibre Channel. In addition, the PCS generates the carrier sense and collision detection indications and it manages the auto-negotiation process by which the network interface (NIC) communicates with the network to determine the network data speed (10,100 or 1000 Mb/s) and mode of operation (half-duplex or full-duplex).
- The PMA sublayer provides a medium-independent means for the PCS to support various serial bit-oriented physical media. This sublayer serializes code groups for transmission and deserializes bits received from the medium into code groups.
- The PMD sublayer maps the physical medium to PCS, and it defines the physical layer signaling for the various media it supports. PMD also includes the *medium dependent interface* (MDI), which is the actual physical layer interface that also defines the actual physical attachment for different media types such as connectors.

The GbE defines two different bit rates. Raw data is formatted at the MAC layer and is passed via the GMII over to the physical layer at 1000 Mb/s. This is known as *instantaneous transmission rate for encoded MAC data*. However, at the physical layer the 8B/10B coding increases by 25% the line bit rate to 1.25 Gb/s. This is known as the *instantaneous transmission rate*.

7.2.2 10 Gigabit Ethernet

The success of the GbE led to a more advanced Ethernet protocol to match the 10 Gb/s (OC-192) data rate of SONET/SDH and to support 6–8 million ports. However, because of the very high data rate, the 10GbE supports transmission over fiber-optic physical media, multi-mode fiber (MMF) and single-mode fiber (SMF) and for fiber lengths up to 10 km and in certain cases up to 40 km (IEEE 802.3ae); in addition, a variant of it was defined for copper medium (IEEE 802.3ak).

The 10GbE enables local area networks (LAN), metropolitan area networks (MAN), wide area networks (WAN) and storage area network (SAN), to operate in full-duplex only and without CSMA/CD. In addition, 10GbE compatibility with OC-192 SONET (10 Gb/s) is accomplished through the definition of a *WAN interface sublayer* (WIS). Thus, according to IEEE 802.3ae:

- 10GBASE-LX4 (L stands for long range) supports SMF transmission up to 10 km at 1300 nm window (O-band: 1260–1360 nm) using a coarse wavelength division multiplexing (CWDM) grid. It also supports MMF at 1310 nm for fiber lengths up to 300 meters. (In addition, ITU-T G.694.2 specifies a CWDM grid within the range 1270–1610 nm with 20 nm channel spacing using water-free SMF). This version may also be suitable to FSO transmission.
- 10GBASE-EX4 (E stands for extended long-range) supports SMF transmission up to 40 km in the 1550 nm window (C-band). This version may also be suitable to FSO transmission.
- 10GBASE-SX4 (S stands for short range) supports MMF transmission and fiber length up to 550 meters powered by 850 nm VCSEL lasers. The 10GbE MMF length is specified for the 850 nm channel bandwidth as follows: up to 82 meters for 500 MHz.km for 50 µm core fiber, 66 meters for 400 MHz.km for 50 µm core fiber, 33 meters for 200 MHz.km for 50 µm core fiber, and 26 meters for 160 MHz.km for 62.5 µm core fiber. In addition, the fiber length may be extended to 300 meters (and perhaps up to 1 km) with the 850-nm 50-µm optimized fiber. Currently, 10 Gb/s over MMF with 850 nm VCSEL lasers is a cost-effective solution in access networks compared with SMF solutions, which may be twice as expensive. This version may also be suitable to FSO transmission.
- 10GBASE-CX4 supports *twinax* copper cable up to 15 meters for datacenter applications.

The actual 10GbE bit rate on the medium is higher than 10 Gb/s; this is the result of 8B/10B or 64B/66B coding; that is, a 25% increase in the actual bit rate. Thus, 10GbE-4 transmits at 4×3.125 Gb/s (8B/10B) over four optical channels (that is, 3.125 Gb/s/channel), called *lanes*, with a coarse channel spacing of 24.5 nm. The nominal wavelengths defined for each lane are:

Lane 0: 1275.7 nm
Lane 1: 1300.2 nm
Lane 2: 1324.7 nm
Lane 3: 1349.2 nm

The reasons for transmitting over four coarsely spaced wavelengths are:

- Inexpensive lasers that are directly modulated and without cooling
- No need for dispersion compensation and equalization for lengths of 10 to 40 km at such low bit-rate
- Polarization effects are negligible
- Optical devices have lower cost and lower maintenance.

Based on this, the 10GbE MAC layer instead of producing a single serial data stream (like the GbE) it maps on four parallel lanes the byte-organized data in a round-robin manner. For example, on the transmit path, the first byte aligns to Lane 0, the second byte to Lane 1, the third byte to Lane 2, the fourth byte to Lane 3, the fifth byte to Lane 0, and so on. In addition, in order to avoid ambiguity of finding the start of the Ethernet frame on a lane and to facilitate frame synchronization, the 10GbE places the start control character of a frame on Lane 0. This is accomplished by adjusting the typical 12-byte idle inter-packet gap (IPG) length either by padding to 15 bytes or by shrinking it to 8–11 idle bytes. A third option uses an averaging method; it includes a deficit idle counter that keeps track of the number of idle bytes added to or subtracted from 12 (ranging from 0 to 3) so that over many frames the 12-byte average is maintained.

7.3 THE TCP/IP PROTOCOL

The Internet protocol (IP), coincident with the evolution of the personal computers and other communication protocols and technology, enable the user to communicate and to transfer files over the network almost instantaneously. Since then, the Internet had a fantastic spread in the business as well the residential markets, creating an appetite for better and more services that could not be anticipated before. Thus, from the initial version of IP, the IPv4 (IP version 4) standard came out, which was quickly replaced with a more potent, the IPv6 (IP version 6) to address issues the its predecessor had not, particularly security and address space.

IPv6, defines an extended header with address space of 128 bits (that is, 2^{128} Internet addresses), dynamic assignment of addresses, improved options, improved scalability, new traffic mechanisms, packet labeling for better defined traffic flow, real-time any-cast services, and it includes authentication and secure encapsulation. That is, IPv6 emulates the robustness and inherent security of the synchronous network and in fact it threatens traditional voice and video services with new Internet offered services such as *Voice over the Internet protocol* (VoIP) and *Video over the Internet protocol* (Video-o-IP).

7.3.1 The Transmission Control Protocol

The Transmission Control Protocol (TCP) is a connection-oriented protocol, by which a connection is set-up by defining parameters. It is a transport-layer protocol that is built over the IP and provides congestion control mechanisms by using the sliding window scheme; the *round trip time* (RTT) delay is used as a measure to define the sliding window length.

The TCP receives data from the application layer, which it segments and forms packets by adding its own overhead octets. The overhead contains the following fields:

- The *source port* (2 octets) indicates the sending user's port number
- The *destination port* (2 octets) indicates the receiving user's port number
- The *sequence number* (4 octets) able to number $2^{32}-1$ frames
- The *acknowledgment* number (4 octets)

- The *header length* (4 bits) indicates the length of header counted in 32-bit words
- The *reserved* (4 bits)
- The *urgent* (1 bit) indicates whether the urgent-pointer is applicable or not
- The *acknowledgment* (1 bit) verifies the validity of an acknowledgment
- The *push* (1 bit) indicates, when set, that the receiver should immediately forward the frame to the destination application.
- The *reset* (1 bit) instructs, when set, the receiver to abort the connection
- The *synchronize* (1 bit) is used to synchronize the sequence numbering
- The *finished* (1 bit) indicates that the sender has finished transmitting data
- Not used (2 bits)
- The *window size* (2 octets) specifies the window size
- The *checksum* (2 octets) checks the validity of the received packet
- The urgent pointer (2 octets) directs the receiver, when set, to add up the value of the pointer field and the value of the sequence number field in order to find the last byte of the data and to deliver urgently to the destination application
- The *options* (up to 4 octets) specifies functions not available in the basic header
- The padding (1 octet) concludes the TCP header. The data field is appended at the padding field.

This comprehensive TCP header is better suited to applications that require reliability but not speed such as e-mail, wide world web, file transfer, remote terminal access and mobile.

7.3.2   The User Datagram Protocol

The User Datagram Protocol (UDP) is a transport-layer connectionless protocol. UDP over IP provides the ability to check for the integrity of packet flow. UDP provides error control but not as efficiently as the TCP. Like TCP, so UDP adds a header to a segment of data. The various fields added to a data segment are as follows; notice the absence of any acknowledgment fields in the UDP header, since this is a connectionless protocol:

- The *source port* (2 octets) indicates the sending user's port number
- The *destination port* (2 octets) indicates the receiving user's port number
- The *UDP length* (2 octets) specifies the length of he UDP segment
- The *UDP checksum* (2 octets) contains the computed checksum, and this field concludes the TCP header. The data field is appended at the UDP checksum field.

The UDP protocol was developed with header simplicity, as compared with the TCP header, in order to provide faster and more efficient delivery. Thus, the UDP simplicity, although not as reliable as the TCP is better suited to applications requiring short delays.

7.3.3 The Real Time Transport Protocol

The Real Time Transport (RTP) protocol provides the basic functions required for real-time applications. RTP segments data into smaller application data units (ADU), to which it adds its own header to form application level frames to run over transport protocols.

RTP is applicable to real time applications that can tolerate a certain amount of packet loss, such as voice and video. However, RTP includes a mechanism that notifies the source of the received quality of packets so that the source may make some rate adaptation to improve transmission and throughput quality or packet delivery quality.

The RTP packet format contains a header (16 octets), the datagram, and the contributing source identifier (4 octets) as a trailer. The added overhead by RTP is as follows:

- The *version* (2 bits) indicates the protocol version.
- The *padding* (1 bit) indicates that there is a padding field at the end of the payload.
- The *extension* (1 bit) indicates the use of an extended header.
- The *contributing source count* (4 bits) indicates the number of contributing source identifiers.
- The *marker* (1 bit) marks a boundary in a data stream. In video applications this bit is set to denote the end of a video frame.
- The *payload type* (7 bits) specifies the payload in the RTP frame. It also contains information about encryption or compression.
- The *sequence number* (2 octets) identifies the numbered sequence of packets after segmenting the data stream.
- The *timestamp* (4 octets) indicates the time the first byte of data in the payload was generated.
- The *synchronization source identifier* (4 octets) identifies the RTP source in a session.
- The contributing source identifier (4 octets) is an optional field, placed after the datagram, and it indicates the contributing source(s) of the data.

A more comprehensive real time protocol is the *Real Time Control Protocol* (RTCP) that runs on top of UDP and it supports enhanced performance functions by using multicasting to provide performance feedback. This is supported by defining several types of RTCP packets, such as the *sender reports*, the *receiver reports*, the *source descriptor*, the *goodbye*, and other *application-specific* types.

7.3.4 The Internet Protocol

The Internet protocol (IP) is a packet technology defined as *best-effort connectionless* [6]. Thus, IP does not establish a fixed (switched) path dedicated for the duration of a session, but instead, it assembles packets of variable length, and by store-and-forward it delivers each one over one or more routes, taking advantage of the temporal avail-ability of bandwidth resources throughout the network.

At the receiving end, because packets arrive with different latency and perhaps out of sequence, the packets include in their overhead a source and a destination address, a packet sequence number, and error control and other information. Thus, although the IP network did not measure up in real time delivery compared with the circuit switched network, it proved to be an economical method for delivering data and it was rapidly deployed. For example, latency exceeded 500 ms, whereas the acceptable round trip delay in synchronous networks is <300 ms (round trip delay is measured from phone-to-phone and back), quality of service (QoS) was limited to best effort whereas the circuit switched network has better than 99% availability, and privacy was not part of the initial IP strategy.

The IETF Internet Protocol Performance Metrics group (IETF IPPM) and the cooperative association for Internet data analysis group (CAIDA) has defined IP per-formance metrics for evaluating a data network. Among these metrics are:

- Symmetry or asymmetry characteristics of data flow (geographical, temporal, and protocol related),
- Packet length distribution (predominant packet size on the IP network)
- Length of packet train or packet flow distribution (that is, the typical number of packets in a single transaction)
- Causes of packet delay
- Causes of network traffic congestion
- Protocols and their application on the IP network (TCP, http)

The explosion of IP demanded added updates (IPv4, IPv6) to support *voice over IP* and real-time compressed *video over IP* from multiple sources. However, since all sources are not equal, there is a *trust rate* assigned to each. The highest trust rate is assigned to *directly connected interfaces* or *manually entered static routing* and the lowest trust rate to *internal border gateway protocol*. That is, the highest trusted rate sources are similar to traditional synchronous communications networks. As a result, in an effort to offer QoS, service flexibility, granularity, scalability, shared-access, point-to-multipoint, customized bit-rates, and easy service migration in optical networks the Internet protocol is encapsulated in "several forms", such as "Internet over SONET", "Internet over ATM over SONET", and so on.

7.4 THE ATM PROTOCOL

Asynchronous Transfer Mode (ATM) is a data protocol designed to provide quality of service (QoS), service type flexibility, semantic transparency, maintenance and reliability.

The ATM frame (called *cell*) has a fixed short length of 53 octets, Figure 7.4; 5 octets for header and 48 octets for data.

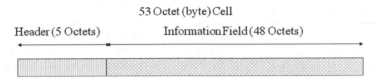

Figure 7.4. The ATM fixed-length cell consists of 53 octets, 5 for the header and 48 for the information field.

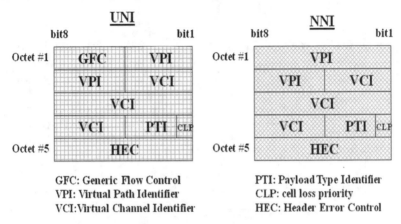

GFC: Generic Flow Control   PTI: Payload Type Identifier
VPI: Virtual Path Identifier   CLP: cell loss priority
VCI: Virtual Channel Identifier   HEC: Header Error Control

Figure 7.5. The ATM header id defined differently for the UNI and for the NNI.

The definition of the ATM cell header at the user to network interface (UNI) differs from that at the network to network interface (NNI), Figure 7.5.

The 4-bit GFC field is defined at the UNI only (at the network access) to assist the control of cell flow but NOT for traffic flow control. GFG is not carried passed the UNI throughout the network; that is, at the NNI the GFC field is not defined.

The VPI/VCI field consists of 24 bits. They are labels identifying a particular virtual path (VP) and virtual channel (VC) on a link. The switching node uses this information and with routing information (tables) that has been established at connection set-up, it routes cells to the appropriate output ports. The switching node changes the input value of the VPI/VCI fields to new output values.

The PTI field consists of 3 bits. It identifies the payload type and it indicates congestion state.

The CLP field consists of 1 bit; 0 = high priority and 1 = low priority. It is set by the user or service provider. Under congestion, the CLP status determines whether cells will be dropped or pass.

The HEC field is used for error detection/correction. This code detects and corrects a single error in the header field or it detects multi errors. It is based on the $x^8 + x^2 + x + 1$ CRC code.

The HEC is also used for cell delineation. The remainder of the polynomial is EX-ORed with the fixed pattern "01010101", and this is placed in the HEC field. At

TABLE 7.1. Definition of ATM cells

Idle Cell:	It is inserted by the physical layer in order to adapt the cell rate to the available rate of the transmission system
Valid Cell:	It is a cell with no header errors, or with a corrected error
Invalid Cell:	It is a cell with a non-correctable header error
Assigned Cell:	It is a valid cell that provides a service to an application using the ATM-layer service
Unassigned Cell:	This is not an assigned cell

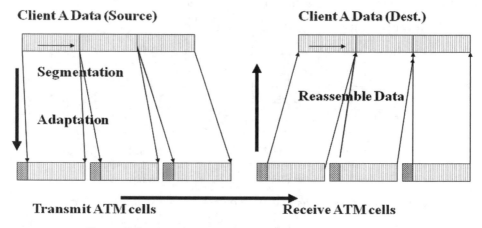

Figure 7.6. The Segmentation and Reassembly (SAR) process.

the receiving end, the generating polynomial results in a HEC pattern "01010101" which is used to locate the start of the cell.

Some ATM cells carry client data and some are defined for other usages. A list of such cells is, Table 7.1:

Client data may consists of long packets (much longer than 53 bytes) or by a continuous bit stream. To transport client data over ATM, data is segmented up to 47 or 48 octets, overhead is added and a string of ATM cells is formed. This function is known as *segmentation and reassembly* (SAR), Figure 7.6.

ATM technology defines five adaptation layers (AAL) each suited for different payload type and services such as, voice, video, TCP/IP, Ethernet, and so on. AAL-5 is straight-forward and more efficient for point-to-point ATM links. AAL-5 cells are transmitted sequentially and thus there is no mis-sequencing protection, whereas error control is included in the last ATM cell (SAR_PDU). As an example, Figure 7.7 illustrates the process for TCP/IP packets over ATM and the corresponding OSI layers.

ATM defines extensive traffic parameter requirements and quality of service features. *Quality of Service* (QoS) of a connection relates to *cell loss, cell delay,* and *cell delay variation* (CDV) for that connection in the ATM network. CDV is a performance metrics pertaining to path speed and is defined as the variability of cell arrival for a

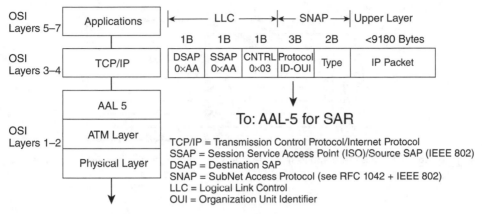

Figure 7.7. ATM adaptation layer-5 process for the TCP/IP protocol.

given connection.Traffic shaping (TS) is defined as the function that alters the flow (or rate) of cells in a connection, to comply with the agreed QoS requirements (rate reduction, cell discarding).

The ATM service level agreement (SLA) includes parameters such as:

- *Peak cell rate* (PCR), is defined as the permitted burst profile of traffic associated with each UNI connection. This is the maximum cell rate in a time interval. In addition, there is
- *Sustainable cell rate* (SCR) is defined as the permitted upper bound on the average rate for each UNI connection.
- *Burst tolerance* (BT) defines the tolerance on additional traffic above the SCR. When this tolerance is exceeded, the ATM traffic is tagged as excessive traffic and it may be lost.
- *Maximum burst size* (MBS) defines the length of bursty cells in a period.

Semantic transparency determines the capability of a network to transport information from source to destination with an acceptable number of errors and performance metrics. Errors are defined on the bit level as well as on the packet level, such as the bit error rate (BER), and the packet error rate (PER).

ATM defines a large variety of services, such as Constant Bit Rate (CBR), Variable Bit Rate (VBR), Real time Variable Bit Rate (rt-VBR), Non-real time Variable Bit Rate (nrt-VBR), Available Bit Rate (ABR), and Undefined Bit Rate (UBR).

ATM cells are transported over an ATM data network that consists of switching nodes capable to switch ATM cells; consists of edge nodes that provide the policing function, a function that verifies that the service level agreement by the client and the service provider is met; and flow control functionality that locates traffic congestion areas and activates mechanisms to avoid them or to guarantee delivery of high priority traffic. During the connection admission control (CAC) process at UNI, the traffic parameters and the requested ATM services are provided by the user to the ATM node.

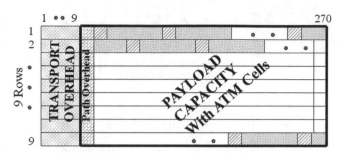

Figure 7.8. ATM cells are mapped in a concatenated manner to fill the SONET/SDH payload envelope.

An end-to-end connection is established with a series of *virtual channel (VC)* links. This is known as *Virtual Channel Connection* (VCC). Each switching node, upon receiving a cell and based on routing translation tables, translates the incoming VCI in the ATM header cell into an outgoing VCI. Thus, the VCI value is different from node to node. The translation values at each node are determined during the set-up of the connection.

An end-to-end connection of a bundle of virtual channel links is established from source to destination and it is known as *Virtual Path Connection* (VPC). The VPI in the ATM header cell identifies the VPC. Each ATM cell in the bundle has the same VPI. Each switching node, upon receipt of a bundle and based on routing translation tables, translates the incoming VPI into an outgoing VPI. Thus, the VPI value is different from node to node. The translation values at each node are determined during the set-up of the connection.

VPCs may be permanent or on demand. Permanent connection is established by node provisioning at the subscription phase and thus signaling procedures are not needed. Connections on demand require signaling procedures so that VPCs may be set-up and released by the network or by the customer.

ATM cells may also be transported over SONET/SDH. ATM cells are mapped in concatenated (STS-Nc) payloads. For example, in STS-3c, ATM cells are mapped in the payload capacity by aligning the byte structure of every cell with the STS-3c byte structure. The entire payload capacity (260 columns) is filled with cells, yielding a transfer ATM capacity of 149.760 Mb/s, Figure 7.8. In STS-12c, ATM cells are mapped into the payload capacity by aligning the byte structure of every cell with the STS-12c byte structure. The entire payload capacity (1040 cols) is filled with cells, yielding a transfer capacity of 599.040 Mb/s. If client cells are not available during the filling process, idle ATM cells are used.

7.5 WIRELESS PROTOCOLS

Although FSO is an ultra-bandwidth (or super broad-band) optical technology, it requires line of sight. In many access applications, however, ultra-bandwidth is not as

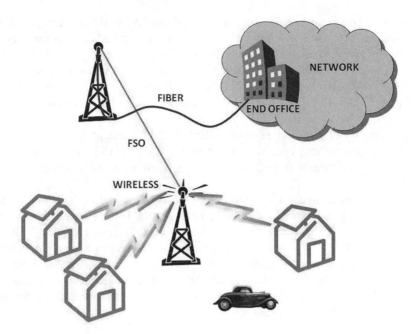

Figure 7.9. FSO may be deployed as backhaul network with a wireless technology in the access network to allow for mobility and for multicasting.

critical of a feature as mobility and multicasting. In this case, FSO may be used as a backhaul network and a wireless technology in the access network, simultaneously multicasting to many end users, shown in Figure 7.9.

In this section, we briefly describe three popular wireless technologies, the WiFi, WiMAX and its rival LTE.

7.5.1 Wi-Fi

Wi-Fi echoes the well known radio term *Hi-Fi* (High Fidelity) and it suggests *Wireless Fidelity*. Nevertheless, Wi-Fi may not mean anything at all and is not a technical term; it was defined by the Wi-Fi Alliance and the technical term is *IEEE 802.11*, after the standard that describes it; "wireless local area network (WLAN) based on IEEE 802.11" describes its applicability and standard.

Wi-Fi is designed to support wireless data transmission at moderate data rates and moderate-to-short distances for personal computers (PC) and laptops, peripheral devices, video game consoles, MP3 players, smartphones, and other devices. Over time, Wi-Fi evolved to support additional services and devices, such as Wireless Voice over the Internet (WVoIP), PC-to-PC direct communication, and the Super Wi-Fi coined by the United States FCC (Federal Communications Commission) planned for longer-distance wireless Internet connections [7].

Wi-Fi is defined for indoor and outdoor applications; a common indoor application is the digital subscriber lines (DSL) with Wi-Fi for wireless data (Internet Protocol) transmission. As a result, Wi-Fi becomes ubiquitous and university campuses, hotel complexes, business facilities, airports, and cities have installed campus and city-wide Wi-Fi access points to provide wireless data services.

Wi-Fi includes multimedia, power-saving, and security features, such as the *Extensible Authentication Protocol* (EAP), the *Wi-Fi Protected Access* (WPA and WPA2 after IEEE 802.11i), and more [8–11], the details of which are beyond the scope of this book.

Although Wi-Fi is described by the IEEE 802.11 standard and its variants, the spectrum assignment and operation is not uniform worldwide. For example, in the United States, 13 channels in the 2.4 GHz band are used, whereas in Japan, 14 channels are used.

Wi-Fi has an *equivalent isotropically radiated power* (EIRP), which is about 20 dBm (100 mW) and thus it has a limited range; in the United States, the maximum power a Wi-Fi device can transmit is limited by FCC Section 15.249 [12]. Thus, a typical 802.11b or 802.11g wireless router with a stock antenna has an outdoor range of about 100 meters, or about 200 meters if an 802.11n router is used. However, if Wi-Fi extenders (acting as repeaters) or directional antennas are used, the distance can be improved up to ten-fold. As a consequence, mobility is confined within the operational range of the Wi-Fi application.

7.5.2   WiMAX

WiMAX stands for *Worldwide Interoperability for Microwave Access*; it is a wireless telecommunications technology and protocol that was defined by the WiMAX Forum for fixed and mobile wireless broadband access services (the so called first/last mile), as an alternative to cable and DSL. It is similar to WiFi but transports much higher data rates, over much longer links, and to many more users.

WiMAX is based on the IEEE 802.16 protocol and was initially defined for data rates up to 40 Mb/s; subsequent protocol variants, such as the IEEE 802.16m, extended the date rate up to 1 Gb/s [13–18]; the variant 802.16d-2004 or "Fixed WiMAX" does not support mobility, whereas the variant 802.16e-2005 or "Mobile WiMAX" supports mobility and other features. Since its definition, WiMAX received worldwide acceptance.

A WiMAX system consists of two parts:

- A WiMAX tower station connected to the Internet via a high-bandwidth medium. This tower station is either connected via a microwave link (or FSO link) to a broadcasting tower antenna (known as backhaul), or itself broadcasts over an area of 8,000 square km; note that, in the backhaul case, line of sight is required.
- A WiMAX receiver and antenna, small to fit in the user's laptop.

According to IEEE 802.16e-2005 protocol, the carrier spacing is constant across different channel bandwidths multiples of 1.25 MHz (typically 1.25 MHz, 5 MHz, 10 MHz or 20 MHz) to improve the spectrum efficiency in wide channels, and to reduce the cost in narrow channels; this is known as Scalable OFDMA (SOFDMA). In addition, bands are not multiples of 1.25 MHz and the carrier spacing is not exactly the same. Carrier spacing is 10.94 kHz. Thus, SOFDMA is not compatible with OFDMA and thus corresponding equipment are not interoperable.

Although there is no globally uniform licensed spectrum for WiMAX (for example, in the US 2.5 GHz is used, whereas in Asia 2.3, 2.5, 3.3 and 3.5 GHz), the WiMAX Forum has published three licensed spectrum profiles, 2.3 GHz, 2.5 GHz and 3.5 GHz. The spectral efficiency of WiMAX is about 3.7 bits per Hertz. Although WiMAX delivers high data rate, like all other wireless technologies, the delivered data rate is inversely proportional to link length.

WiMAX is specified for a range of 50 km radius from the base station (the broadcasting tower antenna), and (depending on standard) at least 70 Mb/s.

Connecting to the WiMAX network is accomplished with various subscriber unit (SU) devices, such as portable handsets, USB dongles, cards or devices embedded in laptops. In addition, WiMAX gateways are available for both indoors and outdoors, which are typically located at the customer premises allowing for connectivity to multiple Internet devices, voice over IP (VoIP), Ethernet devices, and simultaneous voice and data.

WiMAX uses a scheduling algorithm based on which the subscriber station competes only once to gain access into the network. When the network grants access to the subscriber station, the base station allocates to it a time slot, which cannot be used by other subscribers.

WiMAX is more suitable to wireless metropolitan area network (MAN) applications than to wireless local area network (WLAN) or wireless personal area network (WPAN); a wireless technology such as Bluetooth (IEEE 802.15) or wireless USB is better suited to WPAN, and WiFi (IEEE 802.11) to WLAN.

Besides WiMAX, new generations of cellular standards have appeared, known as International Mobile Telecommunications-2000 (IMT—2000), 3G (3rd generation) and 4G (4th generation), the standard of which is issued by the International Telecommunications Union for Radio (ITU-R) [ITU-R M.1457, 1999], and [19]; the interested reader may dwell on this subject in the corresponding standards.

7.5.3 Comparison of WiMAX with Wi-Fi

Both Wi-Fi and WiMAX are wireless technologies for data communications. However, the amount of data rate, the distance and the services each is designed for are quite different, Table 7.2. For example:

- WiMAX is designed for long range (kilometers), whereas WiFi is designed for shorter range (hundreds of meters).
- WiMAX and Wi-Fi are considered non-competing and complementary. For example, WiMAX may be used to provide subscriber connectivity to metropolitan network and Wi-Fi to provide wireless connectivity to a multitude of end user devices within a home, business, campus, and so on.
- WiMAX uses both the licensed and the unlicensed spectrum, whereas Wi-Fi uses the unlicensed spectrum.

TABLE 7.2. Comparison of WiMAX with other wireless LAN technologies

Parameters	IEEE 802.16d-2004 Fixed WiMAX	IEEE 802.16e-2005 Mobile WiMAX	IEEE 802.11 WLAN	IEEE 802.15.1 (Bluetooth)
Frequency Band (GHz)	2–66	2–11	2.4–5.8	2.4
Range	~50 Km	~50 Km	~100 meters	~10 meters
Max Data rate	~134 Mb/s	~15 Mb/s	~55 Mb/s	~3 Mb/s
No. of Users	Thousands	Thousands	Dozens	Dozens

- WiMAX is more expensive than Wi-Fi and thus Wi-Fi is more popular for end-user Internet services.
- WiMAX runs on connection-oriented MAC protocol, whereas Wi-Fi runs on the media access control (MAC) carrier sense multiple access with collision avoidance CSMA/CA protocol.
- WiMAX uses a quality of service (QoS) mechanism based on specific scheduling algorithms, whereas Wi-Fi uses contention access (end users compete for accessing the wireless access point on a random interrupt basis.
- WiMAX supports communication between end-user devices and an access point (AP) only when the end-user device is in range of the base station; Wi-Fi supports direct ad-hoc networking as well as peer-to-peer networking between end-user devices without an access point (AP).
- Both WiMAX and Wi-Fi are standardized protocols; 802.16 and its variants describe WiMAX and 802.11 and its variants describe Wi-Fi.

7.5.4 LTE

The *Long Term Evolution* (LTE) is a set of standards for predominantly mobile communications introduced by the Third Generation Partnership (3GPP) as a natural evolution of the previous cellular technologies Third Generation (3G) and the GSM/UMTS (Global System for Mobile/Universal Mobile Telecommunications System). In late 2009, a proposal was submitted to ITU as a candidate of the "beyond the third generation" (3G) mobile network, namely pre-4G, that had corrections and fixes of the 3G. Since then, the pre-4G has been commercially deployed by certain mobile network and services providers (such as AT&T and Verizon); subsequent advancements culminated in the *LTE Advanced*, also known as the *Fourth Generation* (4G) system, which is defined by ITU-R as follows:

- Scalable data rate and peak data rate up to 1 Gb/s,
- Improved performance at the cell edge,
- Faster switching times between power states,

- Support macrocells (100 km radius) in rural areas, as well as small and smaller cells (tens of meters) in inner-city areas, called *picocells* and *femtocells*, for increased capacity.
- Support much lower power relay nodes,
- Service at least 200 active users in every 5 MHz cell.
- Support high mobility; depending on frequency band, 350 km/h and up to 500 km/h.
- Support an all IP network with sub-5 ms latency (for small packets), and all user types.
- Single antennas as well as multiple-input and multiple-output (MIMO), 4×4 and 2×2, antennas utilizing 20 MHz of spectrum; MIMO is a smart antenna technology that improves channel communication performance.
- Up-link and Down-link (UL/DL) coordinated MIMO.
- Heterogeneous networks, including nomadic and LANs.
- Co-existance with legacy standards; user can call or transfer data using LTE or its predecessor technology 3G/GSM/UMTS.
- Support Mobile-TV broadcast services (Multicast Broadcast Single Frequency Network (MBSFN)).

In addition, LTE uses scalable and flexible bandwidth that exceeds 20 MHz and up to 100 MHz, enhanced precoding, forward error correction (FEC) for improved data performance at longer distances, and improved communication security (a summary of LTE features is found in TR36.912) [20–23].

Because different countries allocate different frequency bands, the LTE is planned to be frequency-band adaptable. For example, it is planned in the bands 700 MHz in North America, 900, 1800, 2600 MHz in Europe, 1800 and 2600 MHz in Asia, and 1800 MHz in Australia. Note that the worldwide deployed GSM operates at the quad-band 850, 900, 1800 and 1900 MHz and the worldwide deployed UMTS on 14 different bands.

7.5.5 Comparison of LTE with WiMAX

LTE is viewed as a competing technology to mobile WiMAX. Both wireless technologies, WiMAX and LTE, are intended to offer broadband mobile services at many megabits per second, and major network and service providers are in a race offering mobile services based on one or the other technology. For example:

- WiMAX is an IEEE specification (802.16e) designed to support mobile Internet Protocol (IP) as high as 12 Mb/s data-rate using Orthogonal Frequency Division Multiple Access (OFDMA) (see previous sections in this Chapter). In addition, the enhanced WiMAX standard (802.16m) is expected to be much faster than its predecessor, capable of delivering up to 128 Mb/s per channel downlink and 56 Mb/s uplink, and to supporting MIMO smart antenna technology. Some major

network and service providers, such as Sprint and Clearwire, have already rolled
out deployment plans offering services based on WiMAX, with 120 million POPs
in all major US markets by 2010.

• LTE was developed as the natural progression of High-Speed Packet Access
  (HSPA), the GSM technology used by some other major carriers, such as Verizon
  and AT&T, to deliver post-3G mobile broadband services, LTE is capable of
  delivering up to 100 Mb/s per channel downlink and at least 50 MB/s uplink, to
  support both frequency division duplexing (FDD) and time division duplexing
  (TDD), and to perform as well as the wired broadband technology does. The
  supporters of LTE claim that LTE provides a more natural upgrade for their
  GSM/UMTS/HSPA/CDMA-based networks and subscribers, and also tha GSM
  is already a dominant mobile standard with more the 3 billion customers worldwide, as of February 2010.

7.6 THE NEXT GENERATION SONET/SDH PROTOCOL

Prior to describing the next-generation SONET/SDH, it is pedagogical to provide a
very brief overview of the legacy SONET/SDH.

7.6.1 The Legacy SONET/SDH

In the 1980s, a new standardized protocol was introduced that defined the specifications
of interfaces, architecture and features of a synchronous optical network, which in the
United States was called SONET and in Europe and elsewhere synchronous digital
hierarchy (SDH); SONET and SDH were defined with enough differences to have two
different standards; SONET in the United States managed by Telcordia (previously
Bellcore) and SDH by ITU [24–38]. Since its introduction, the SONET/SDH network
performed beyond expectation, was quickly adopted by most advanced countries, and
became the de-facto standard of optical networks.

The set of SONET standard interfaces is the *synchronous transport signal level-N*
(STS-N), where N = 1, 3, 12, 48, 192 and 768. The STS-N rate on the optical medium
is known as *optical carrier-N* (OC-N); STS-N indicates the bit rate of the electronic
signal before the optical transmitter. The topology of the network is typically a protected
ring with add-drop multiplexing (ADM) nodes or protected point-to-point with ADMs;
network nodes are known as *network elements* (NE).

Similarly, the SDH set of standard interfaces is the *synchronous transport module
level-M* (STM-M) where M = 1, 4, 16, 64 and 256.

Both SONET and SDH define all layers, from the Physical to the Application. The
physical transmission medium of both SONET and SDH is single mode fiber (SMF).

Both SONET and SDH carry all synchronous broadband rates (DS-n, E-n), asynchronous data (ATM), and well known protocols (Internet, Ethernet, Frame Relay)
encapsulated in ATM first and then mapped onto SONET/SDH.

For maintenance, operations, administration and management, the SONET/SDH
defines three network layers, path, line and section.

- The *path layer* addresses issues related with transport of "services", such as DS3, between path terminating network elements (PTE); that is end-to-end.
- The *line layer* addresses issues related with the reliable transport of path layer payload and its overhead across the physical medium. It provides synchronization and multiplexing for the path layer network based on services provided by the section layer.
- The *section layer* addresses issues related with the transport of STS-N frames across the physical medium and uses the services of the physical layer to form the physical transport. It constructs frames, scrambles payload, monitors errors, and much more.

7.6.2 SONET Frames

SONET/SDH (STS-N) frames come in specific sizes. However, regardless of size, a SONET/SDH frame is always transmitted in 125 µs and because of this each STS-N signal has a specific line rate, Table 7.3.

The SONET/SDH protocol was based on specific size frame-structures; the smallest frame consisted of bytes arranged in a matrix of 90 columns by 9 rows. The first three columns of this frame (STS-1 for SONET and STM-0 for SDH) were allocated for (section and line) *transport overhead*, and the remaining 87 columns were allocated, one for path overhead, 84 for user payload and two columns were unused (called "fixed stuff"), Figure 7.10. The transport overhead consists of two parts, the *section overhead* (row 1 to row 4 and column1 to column 3) and the Line Overhead (row 4 to row 9 and column 1 to column 3) and each octet in them has a specific meaning and function. Similarly, each octet in the path overhead has a specific meaning and function, although the functionality of each octet requires several contiguous frames to comprise the complete functionality of each byte in the overhead.

The payload could be filled by tributary units (TU in SDH) or virtual tributaries (VT in SONET) of specific sizes, that were mapped in groups and then in specific size

TABLE 7.3. Line rates of SONET/SDH frames

Signal Designation			Line Rate
SONET	SDH	Optical	(Mbps)
STS-1	STM-0	OC-1	51.84 (52M)
STS-3	STM-1	OC-3	155.52 (155M)
STS-12	STM-4	OC-12	622.08 (622M)
STS-48	STM-16	OC-48	2,488.32 (2.5G)
STS-192	STM-64	OC-192	9,953.28 (10G)
STS-768	STM-256	OC-768	39,813.12 (40G)

OC-N: Optical Carrier—level N
STS-N: Synchronous Transport Signal—level N
STM-N: Synchronous Transport Module—level N

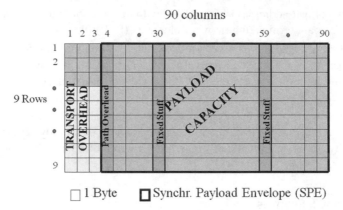

Figure 7.10. The smallest SONET/SDH frame consisted of a matrix of bytes arranged in 90 columns by 9 rows.

payload envelops, using synchronized byte multiplexing. Thus, the upper bound efficiency of an STS-1 is 93.33%. Nevertheless, the actual efficiency is much less (about 60%) because there is wasted bandwidth, as will become evident.

An STS-N frame is transmitted row by row, starting with the first octet (row 1, column 1). When the last octet of the first row is transmitted, it continues with the second row (row 2, column 1) and so on until it reaches the very last octet in the frame (in STS-1, this is row 9, column 90). After the last byte in the frame, the process continues with the first byte of the next frame, and so on.

Since the SONET/SDH frame generated in a network element (NE) may not be in complete synchronism (frequency and phase) with the incoming payload, there is an undetermined phase difference. To minimize latency, SONET/SDH follows the method of dynamic pointer and floating frame that directly maps the incoming payload within the frame, the offset value which is measured and is stored in the overhead of the frame. In addition, SONET/SDH has provisions and mechanisms to justify for frequency variations. Thus, at start-up (or at a reference time t = 0), the offset is calculated, and if the calculated offset remains the same for three consecutive frames, a "no justification" is indicated. If the incoming frequency slightly varies with respect to the local frequency, then positive or negative justification is performed: positive when the incoming rate is a little lower than the node clock, and negative when the incoming rate is a little higher.

7.6.3 Virtual Tributaries and Tributary Units

Although in synchronous communications DS-ns (and E-ns) are tributaries that carry the payload of many customers, SONET defines Virtual Tributaries (VT) and SDH defines tributary units (TU) to carry DS-ns or E-ns. The capacity of a VT depends on the number of octets in it, and because the number of rows is always nine, it depends on the number of columns. Thus, if the number of columns is 3, it is known as VT1.5;

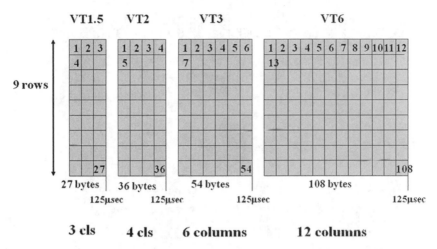

Figure 7.11. Virtual Tributaries (SONET) or Tributary Units (SDH) have specific fixed sizes.

if 4, it is known as VT2; if 6, it is known as VT3; and if 12, it is known as VT6, Figure 7.11. Each VT contains a client signal not necessarily of the same type, and therefore a VT has its own overhead known as VT path layer overhead.

VTs are byte multiplexed to form a group of 12 columns. However, a simple rule applies: a group can only contain the same type of VTs. That is, four VT1.5s, or three VT2s, or two VT6s, and not a mix of them. Figure 7.12 shows the logical multiplexing/demultiplexing hierarchy from/to synchronous TDM to SONET.

Thus, only seven groups fit in a SONET STS-1 SPE, which are byte (or column) multiplexed, with the added path overhead and the two fixed stuff columns. A SONET frame is transmitted within 125 μs.

SDH defines a similar organization and column multiplexing like SONET. However, in SDH, VTs are called tributary units (TU), groups are called tributary unit groups level 2 (TUG-2), and seven TUG-2s are multiplexed in a TUG level 3 (TUG-3), which with the addition of two columns of fixed stuff at the first two columns of the TUG-3 form the SPE. Here, the same rule also applies: a TUG-2 must contain the same type of TUs.

The capacity and data rate equivalent of VTs depends on the number of columns (the number of rows is always 9) which in SONET/SDH are 3, 4, 6 and 12. Because a SONET/SDH frame is transmitted within 125 μs, so is each TU/VT and each byte in a virtual container/tributary; the data rate of each byte in any TU/VT is equivalent to 64 Kb/s; Table 7.4 tabulates the equivalent data rates per VT.

For the various operation, maintenance, administration and control functions the overhead of SONET frames was defined according to line, section and path. The network topology, switch to protection and data-rate objectives were met and perhaps surpassed, and the initially used data rates of up to 622 Mb/s (OC-12) were extended up to OC-768 (40 Gb/s).

Based on the aforementioned description, the legacy SONET/SDH does not have the fine granularity to support modern data payloads and thus the bandwidth efficiency

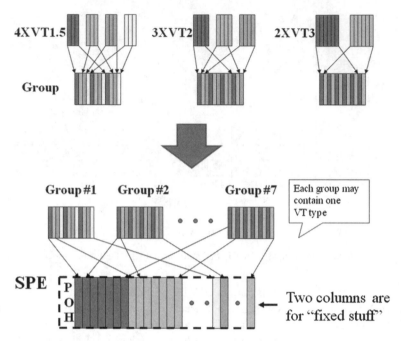

Figure 7.12. Hierarchical and synchronized byte multiplexing of VTs into Groups into the payload.

TABLE 7.4. Equivalent data rate of VTs

SONET	Bit Rate (Mbps)
VT1.5*	1.728
VT2*	2.304
VT3*	3.456
VT6*	6.912

of VT or TU structure and groups is low. Thus, the next-generation intelligent optical network is an evolution of a proven transporting vehicle but reengineered to support a larger variety of data protocols and new services and requirements with cost-efficiency.

7.6.4 STS-N Frames

SONET and SDH define higher capacity frames. For example, a STS-N has N times the amount of columns of a STS-1 (both overhead and payload) but always nine rows. For example, a STS-3 frame has a total of 270 columns, 9 overhead columns, three path overhead and six fix stuff columns. However, if three STS-1s are multiplexed to produce a STS-3, then three overhead pointers must be processed, since each constituent STS-1 may arrive from a different source with different SPE offset, Figure 7.13.

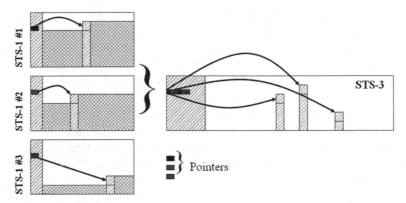

Figure 7.13. Three individual STS-1s, each with own overhead, multiplexed to construct a STS-3 frame.

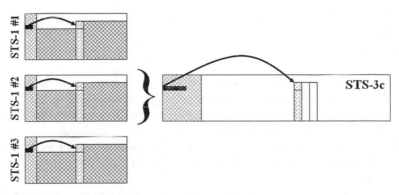

Figure 7.14. Three STS-1s, all with common overhead, multiplexed to construct a STS-3c frame.

7.6.4.1 Concatenation SONET/SDH can also accommodate super large packet payloads that do not fit in a single STS-1 frame. This is addressed by distributing the large packet over N STS-1s and then multiplexed them in a single STS-N, which is denoted as STS-Nc indicating "concatenation". Because the STS-Nc payload has the same origin and destination for all STS-1s in it and all STS-1s have the same frequency and phase relationship among them, there are many redundancies. That is, one pointer processor suffices, the overhead is simplified, one path overhead is needed, and there are fewer fixed stuff columns (the number of columns is calculated by N/3 − 1). Moreover, each node or network element treats a STS-Nc as a single entity and it distinguishes a STS-Nc from a regular STS-N from specific codes written in the unused pointer bytes. Figure 7.14 illustrate the frame of STS-3c.

7.6.4.2 Scrambling SONET/SDH defines a scrambling process so that no long strings of zeroes or ones are present. The scrambler is defined by the generating polynomial: $1 + x^6 + x^7$ which generates a random code 127 bits long. The scrambler is set to 11111111 on the MSB of the byte following the #Nth STS-1 C1 Byte (of a STS-N). Thus, in STS-1, the scrambler starts with the first byte after A1, A2 and C1, and it runs continuously throughout the complete STS-N frame; A1, A2 and C1 are not scrambled because they are used to identify the start of frame.

7.6.4.3 Maintenance The SONET and SDH recommendations define all maintenance aspects, criteria, requirements and procedures to maintain the network element and network operation at an acceptable performance. Requirements include alarm surveillance, performance monitoring (PM), testing, and control features to perform the following tasks:

- Trouble monitoring and detection
- Trouble or repair verification
- Trouble sectionalization
- Trouble isolation
- Testing, and
- Restoration

The network element alarm surveillance takes place at the termination of section and path. This is accomplished by monitoring and processing corresponding bytes in the frame overhead. When a node detects an impairment that occurred in a line, path or a particular VT path, a corresponding alarm indication signal (AIS) is issued. That is, AIS can be on any of the three levels, AIS-L for Line, AIS-P for Path, AIS-V for VT Path. AIS is issued when loss of signal (LOS), loss of frame (LOF) or loss of pointer (LOP) occurs.

SONET performance monitoring (PM) is based on counting code violations in a second, whereas SDH PM is based on counting errored blocks in a second. In the next generation optical networks, the unit of time "second" is very long and may be used for metric and comparison purposes since a rate at 10 Gb/s yields so many bits in a second, which are many more than any packet technology; that is, performance in the next generation optical network must be monitored much faster than what the original SONET/SDH has defined.

Finally, SONET and SDH have the capability to test the signal at different levels by looping back a complete STS-N or individual VTs.

7.7 NEXT GENERATION SONET/SDH NETWORKS

Although the original SONET/SDH was developed for the ring topology, the next generation is applicable to both ring and mesh. The mesh topology has excellent link and service protection, because nodes may be reconfigured to reroute traffic bypassing failures or congested areas. Faults are detected with optical power detectors and performance estimation thresholds. Reconfiguration can be achieved either autonomously or using sophisticated network management procedures that also include traffic balancing and traffic grooming. The next generation synchronous optical network provides a standardized, robust, and efficient method that transports all types of data and packets (IP, Ethernet, Fiber Channel, etc) over SONET/SDH (DoS), in addition to synchronous traditional TDM traffic [39].

7.7.1 Next Generation Ring Networks

The Next Generation Optical Ring (NG-OR) conforms to all previously known ring topologies, unidirectional single fiber, bi-directional single fiber, two-fiber bidirectional, four-fiber bidirectional, and so on. In addition, each wavelength carries data rates at OC-48 (2.5 Gb/s), OC-192 (10 Gb/s), and also OC-768 (40 Gb/s). Some wavelengths (in DWDM) may also carry raw 1GbE or 10GbE.

Ring nodes consist of an optical add/drop multiplexer (OADM) and a network element that disaggregates/aggregates traffic (in the electronic regime). As a consequence, the network elements in the NG-OR support a variety of interfaces that provide aggregation, grooming, and switching capabilities, respond to alarm and error SONET/SDH conditions and support the multiservice provisioning platform (MSPP). Some nodes provide bandwidth and wavelength management and are able to connect two or more NG-ORs supporting the multiservice switching platform (MSSP). Thus, ring nodes receive a diverse client payload of synchronous and asynchronous data and are able to encapsulate in GFP and map in NG-S frames.

The NG-ORs will exhibit advanced fault detection strategies and advanced signaling protocols such as the generalized multi-protocol label switching (GMPLS), which is an evolution of the multi-protocol label switching (MPLS) standard. The key objective is to protect traffic when a wavelength degrades beyond the acceptable threshold or when a fault occurs and to allow secure data to flow over a secure network.

7.7.2 Next Generation Mesh Networks

When several next generation ring networks are connected, a mesh network is constructed known as path protected mesh network (PPMN), Figure 7.15. In this case, the common nodes between two rings are large MSSPs that carry large capacity aggregate traffic from ring to ring.

7.7.2.1 Next Generation Mesh Networks: Protection The PPMN network combines ring and mesh protection strategies. One strategy is based on predetermined redundant paths; for every possible path, an alternate path has been identified. This method allows for the fastest "switch to protection", although the protection path may not be the best possible at the time of failure, as congestion conditions may occur unpredictably over the end-to-end path, and particularly when it crosses subnetworks that are operated by different network operators, Figure 7.16.

Another strategy is based on algorithms (the shortest path, constraint-based, the least congested path or most available path, and others (known from the Generalized Multiprotocol Label Switching (GMPLS)) to identify the best possible path available.

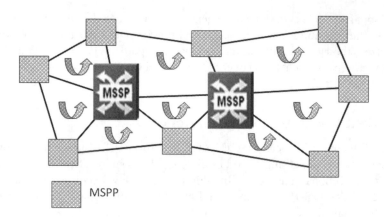

MSPP

Figure 7.15. A mesh network constructed by several interconnected ring networks.

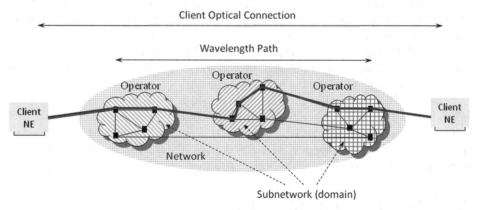

Figure 7.16. The overall network may consists of subnetworks, each operated by different network operators.

These algorithms require knowledge of the status of network nodes, and therefore, they require extensive signaling and complex protocols. Such algorithms are slow in finding the protected path but they do find the best available at the time. Yet another strategy combines quick algorithms that, based on network metrics, identify the best available protection path from a set of predetermined redundant paths.

PPMN resolves multiple failure conditions on the network, on the link and on the channel (wavelength) level. Nodes are provisioned to reroute traffic away from a failure or congestion condition. Faults are detected with power detectors and performance parameter thresholds. Reconfiguration can be autonomous according to SONET/SDH standards and to new wavelength management strategies or it can be with sophisticated multiprotocols that perform traffic balancing and traffic grooming.

7.7.2.2 Next Generation Mesh Networks: Traffic Management   The next generation intelligent optical mesh network consists of network elements and fiber links; in mesh-FSO consider fiberless links. Each link has a maximum capacity calculated from the product (number of wavelengths)X(bit rate per wavelength). However, the effective traffic per fiber link is less than this, as many frames or packets over each wavelength may be either idle or frames for network operations, administration and management (OA&M). Thus, although the network elements may handle the same amount of maximum traffic, bandwidth management deals with balancing the effective traffic per wavelength and per fiber link: fast, without excessive complexity, and cost-efficiently. This is accomplished by monitoring link status and traffic across the network. Monitoring traffic in the Next Generation Network requires intelligence because traffic is dissimilar (voice, video, high speed data of various protocols, such as IP, Ethernet, etc.)

7.7.2.3 Next Generation Mesh Networks: Wavelength Management
When lightpath connectivity is requested, the best route across a mesh network is searched for. Depending on network capability and routing algorithm a wavelength assignment over the best path is made, which it may consist either of a single (and same) wavelength or of concatenated (and different) wavelengths; here, we assume DWDM links. At this point, it is important to identify certain issues related to wavelength assignment.

In the case of the same wavelength assignment over the complete optical path, each optical switching node on the path has been provisioned of the input-output connectivity. That is, each node "knows" the source, the wavelength number and the destination for each wavelength through it. However, in the case of concatenated wavelengths and because wavelength conversion (or translation) is needed at each optical switching node, the next node needs to "know" where the source, the destination, the wavelength number at the input, and the wavelength number at the output. In mesh-FSO networks with a moderate number of nodes, a coarse WDM may allow for wavelengths to be pre-assigned.

In the case where wavelength conversion at nodes is required, then, wavelength conversion is managed with two different strategies, (centralized) network wavelength management and distributed wavelength management.

- In the centralized case, a network wavelength management function provisions each node with *wavelength assignments* establishing semi-static cross-connectivity over selected paths. This case depends on a centralized database and algorithm that finds the optimum shortest and bandwidth efficient path with fewer wavelength conversions. This case also implies that all communications interfaces with the various sub-networks or domains are compatible; this may become an issue when the path extends over a multi-vendor network.
- In the distributed case, there is an additional optical channel that is common to all nodes and over which control messages flow; this is known as *supervisory channel* (SUPV). These messages contain the input-output-wavelength associations, in addition to other management messages. Thus, optimization of wavelength re-assignment is left to each node.

A supervisory channel (a channel dedicated for control messages) conveys messages from node to node very quickly and thus allows for *dynamic node reconfigurability*. *Dynamic system reconfigurability* is also required for system and *network upgrades, scalability* and *service restoration*. Network upgrades entail downloading new software versions and new system configurations. During network scalability and upgrades, service should not be affected.

7.7.2.4 Next Generation Mesh Networks: Network Management   The
next generation intelligent optical network manages the network elements with a simplified management protocol suite. ITU-T has defined the general management functionality known as Fault, Configuration, Accounting, Performance, and Security, or FCAPS. FCAPS is not the responsibility of a specific layer of the TMN architecture but different layers perform portions of FCAPS management functions. For example, as part of fault management, the EML logs in detail each discrete alarm or event, then filters the information and forwards it to an NMS (at the NML layer), which performs alarm correlation across multiple nodes and technologies and root-cause analysis.

7.7.2.5 Next Generation Mesh Networks: Service Restoration   The next
generation all-optical network consists of a multitude of optical and photonic components over the end-to-end path. Consequently, component failure (including fiber cuts), component degradation due to aging, specification degradation due to environmental condition changes (such as temperature and stress), and photon-matter interactions will affect the quality of one or more signals on the path and thus the quality of service.

Among the various degradation contributors are spectral drift and spectral noise, jitter, optical power attenuation and also loss, amplification gain drift, and so on.

Despite the significance each degrading contributor has, the quality of each signal and hence service needs to be monitored and a strategy must be incorporated in each node of the overall network (from access to backbone) to assure that service is delivered at the expected performance level on each link and on the overall end-to-end path. This means that when service is degraded below an acceptable threshold, the network is intelligent enough to perform service restoration either autonomously (in the case of distributed control) or via intelligent centralized control.

Service restoration is the action which either removes the affecting cause or moves the affected channel from one wavelength to another (in WDM networks). The latter action requires certain wavelengths to be reserved for service restoration, and in some cases it may also necessitate wavelength conversion.

We classify three service degradation cases: single channel service degradation, multiple service degradation (over a single fiber), and all-channel service degradation (over a single fiber) [40, 41]. Each degradation classification requires different complexity and restoration strategy.

In traditional SONET/SDH, the signal quality is monitored via the bit error rate (BER) by calculating the BIP-8 byte in the overhead; there is one BIP-8 for the line and one for the section parts of a SONET/SDH link. These calculations are recorded in the error control byte B3 (and B1, B2) of the corresponding overhead, which the next node reads and responds with a received error indication (REI-L, REI-P) if the error rate has crossed the acceptable threshold of 10^{-n}. When an automatic switch to protection (line, path or lightpath) is triggered, SONET/SDH accomplishes this in less than 50 ms. When the path changes, the path overhead byte J1 is also updated.

7.8   NEXT GENERATION PROTOCOLS

The SONET and SDH protocols were defined prior to WDM for the efficient transport of synchronous TDM traffic and some data asynchronous traffic over single-wavelength fiber networks. Mapping synchronous payload was achieved onto VTs (SONET) or VCs (SDH), and transport over fiber used one of two wavelengths, 1310 nm or 1550 nm, or both.

SONET and SDH proved themselves for reliability and bandwidth scalability, and over a ten year period it enjoyed such growth that it indirectly stimulated an exploding growth in data services, and in an unanticipated way, SONET/SDII became a victim of its own success: the exploding growth demanded much more bandwidth and the single-optical channel SONET/SDH was quickly exhausted. Thus, the next-generation SONET/SDH (NG-S) was defined to alleviate the shortcomings of the legacy SONET/SDH. To contrast the legacy with the Next Generation SONET/SDH, a comparison is made of existing versus desired features in Table 7.5.

TABLE 7.5. Comparison between the legacy and the next generation SONET/SDH

	Legacy SONET/SDH	Next Generation SONET/SDH
Topology:	Ring and Point to point only	Many (Ring, Mesh, Point-to-point, Tree)
Bit Rate:	OC-n (predefined)	OC-n and others (increased granularity)
Interfaces:	OC-n	Supports interfaces from DS1 to OC-768
Optical channels:	One per fiber (1300 nm/1550 nm)	Supports DWDM
Payload Efficiency:	Synchronous & ATM mapping; less efficient for packet	Support all payload mappings with high efficiency, encapsulation and concatenation
Switching:	Low order or high order	High order, low order, packet
Concatenation:	Contiguous only	Contiguous and virtual
Reliability:	High	High
Functionality:	SONET/SDH defined	Multiple, integrated in the same NE
Protection strategy:	<50 ms switching to protection but for channel only for ring and point-to-point topology	<50 ms switching to protection for channel, line, and path for many topologies
Cost (Bandwidth-km)	High	Low

Since the first introduction of SONET/SDH, the optical network became more intelligent [42]. The next generation SONET/SDH supports the multi-service provisioning platform (MSPP) and the multi-service switching platform (MSSP).

- The MSPP provides aggregation, grooming, and switching capabilities. It responds to alarm and error SONET/SDH conditions, and it supports new protection schemes for different topologies such as ring, multi-ring, mesh, and point-to-point. The significance of the MSPP is that it is the edge node that interfaces with diverse client payloads, or tributaries (OC-n, GbE, IP, DS-n). When a tributary fails, then the legacy SONET/SDH practices are adopted. That is, when the MSPP node detects a signal failure at its input, it declares loss of signal (LOS) and generates an alarm indication (AIS-L, AIS-P), which is transmitted in all affected virtual containers (VC). Nodes that receive the alarm indication respond with a received defect indication (RDI-L for line and RDI-P for path)
- The MSSP provides bandwidth and wavelength management via large, non-blocking switching fabrics (cross-connects).

7.8.1   Concatenation in NG-S

Traditional SONET/SDH had defined *contiguous concatenation*. The next-generation SONET/SDH has expanded this by defining *virtual concatenation*.

7.8.1.1   *Contiguous Concatenation*   Contiguous concatenation (CC) supports mapping of very long packets that exceed the capacity of the NG-S synchronous payload envelope (SPE). According to this, CC allows mapping of the packet over two or more contiguous SONET/SDH frames. However, the efficiency of mapping with CC is correlated with packet length and data rate. For example, the efficiency of 10 Mb/s Ethernet mapped in contiguous legacy SONET STS-1s or SDH VC-3s is estimated to be only 20%. Similarly, the efficiency for 100 Mb/s Ethernet mapped in contiguous STS-3c or VC-4 is estimated to be 67%, and for 1 Gb/s Ethernet mapped in STS-48c or VC-4-16c is estimated to be 42%.

Although contiguous concatenation may seem simple, the receiving end must be able to recognize the contiguous mapping and extract from the contiguous SPEs the packet(s). Contiguously concatenated frames have simplified section and line overhead, and they require a single path overhead column, Figure 7.17.

7.8.1.2   *Virtual Concatenation*   The idea of virtual concatenation (VC) is imported from the Internet. Thus, a *high-order* (HO) frame or a high data rate (such as GbE) is segmented into *lower-order* (LO) smaller containers or packets, and each LO container or packet is fit in and transported independently by different SONET/SDH payloads over independent separate paths to meet efficiency. Based on this, high data rates that do not "fit" in a SONET/SDH STS-n/STM-m payload envelop are partitioned to "fit" into more than one, hence, "virtual concatenation". For example, 1GbE can be partitioned to "fit" in the payload of two independent STS-12s, called STS-12-1v and

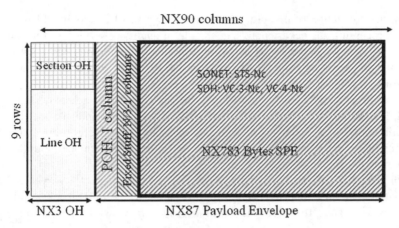

Figure 7.17. Contiguous concatenation of the NG-S has simplified the frame structure and efficiency.

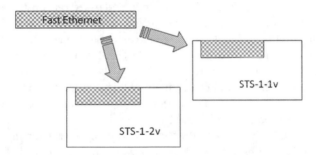

Figure 7.18. Fast Ethernet frames may be mapped onto two NG-S STS-1s.

STS-12-2v. Similarly, 100BASE-T Ethernet may be mapped in two STS-1s, STS-1-1v and STS-1-2v, Figure 7.18.

At a receiving point in the network, the LO containers or packets are collected and reassembled to their original form. A consequence of the separate independent paths is that the two containers may arrive with a *differential delay*, which implies that at the destination the containers must be buffered, arranged in the correct order, realigned, and then reassembled. Thus, although CC solves the problem of fitting very long containers or packets in SONET/SDH contiguous SPEs, VC solves the problem of transporting long-length and very high data rate packets over independent and separate SPEs with increased efficiency.

7.9 THE GMPLS PROTOCOL

7.9.1 Before GMPLS: MPLS

The transport of packetized data (e.g., IP) over a network that supports a different protocol (such as ATM) requires the IP to undergo segmentation and adaptation. In general, the adaptation of one protocol to another protocol results in added overhead, latency, processing and occasionally a mismatch of quality of service, so that the overall network efficiency is decreased. The multiprotocol label switching (MPLS) was defined in an effort to decrease the inefficiency or increase efficiency of multiple data networks that are one on top of the other.

According to MPLS, one or more labels are attached to IP packets when they enter a label edge router (LER) of a MPLS network. These labels indicate the next router destination in the MPLS network and are calculated by a search algorithm and signaling messages that identify and establish the best path; the path from source to destination is known as *tunnel* [43–45].

When a label-switched router (LSR) receives a MPLS packet, it forwards it to one of its outputs which is selected according to the label value in the packet and the port it was received. Thus, the LSR function in the router may swap the label in the packet by another label if the packet is switched to a different output port.

Connections established with the MPLS protocol are called label-switched paths (LSP). Routing protocols determine the LSPs for predefined traffic classes, known as forward equivalent classes (FEC). FECs are specified based on constraints such as QoS parameters, entry port number, and source (originating address). The LSP is defined by the label attached to the packet. Labels are distributed in the MPLS network by a label distribution protocol (LDP). Each MPLS node (LSR router) constructs input-output mapping tables based on which it routes MPLS packets. Thus, a path is defined by a sequence of labels as they have been defined and distributed by LDP.

When a failure or congestion is experienced in a LSP, the MPLS protocol reroutes traffic. This may be accomplished either by pre-established alternative routes (required for time-critical and high priority MPLS packets) or by re-calculating another route, which requires calculations and signaling, and which results in label changes.

When MPLS is over WDM, the optical network control plane needs not only to find the best route available but also to assign and provision a wavelength path. That is, to establish lightpath connectivity over the WDM network. However, with DWDM, there are several different methods to establish lightpath connectivity, which MPLS did not address efficiently.

7.9.2   The GMPLS

A node that supports the Generic Multi-Protocol Label Switching (GMPLS) protocol advertises its bandwidth availability and optical resources (that is, link type, bandwidth, wavelengths, protection type, fiber identifier) to its neighboring nodes. It also requests from its neighbors their own status via signaling messages; this is known as *neighborhood discovery*.

Therefore, the GMPLS runtime must be short and provisioning fast, as well as switch to path protection and restoration.

Fast restoration is particularly important because [46, 47] WDM networks carry traffic at many gigabits per second per wavelength and long restoration time implies an enormous amount of lost packets. Switch to protection is typically according to one of the well-known strategies, $1 + 1$ or 1:N.

GMPLS includes port switching, λ-switching and TDM. It employs search algorithms to find the best path available within a mesh optical network topology (even under impairment conditions), to provide the necessary signaling messages, to establish end-to-end and also link connectivity, topology discovery, connection provisioning, link verification, fault isolation and management, and restoration. These algorithms find the best path that meets the traffic requirements such as required bandwidth, traffic priority, real-time aspects, deliverability, and others, which are familiar to SONET/SDH traffic types and service levels. They also monitor the established path for congestion and failures.

GMPLS appends a calculated label to the packet. This label describes the physical port, the assigned wavelength, and the fiber. If the lightpath enters another node and departs from it without a change in the established lightpath, then the label remains the same. However, if there is a change in the lightpath (due to congestion, traffic balancing or as a result of switch to protection), then the node finds a new (best) path according to an algorithm, sends downstream a RSVP generalized label request or a label request message in CR-LDP, re-calculates a new label and replaces the old one.

A GMPLS node consists of two functions, one that interfaces the client side or the GMPLS aggregator, and another that interfaces the optical WDM network.

The network interface optically multiplexes the received traffic from the aggregator and multiplexes it in the WDM signal. This function is similar to optical add-drop multiplexing with optical cross-connect. The GMPLS aggregator receives client signals, aggregates them, forms labels and packets and hands them to the optical network interface.

To meet quick route discovery and fast response times, GMPLS assumes distributed control and semi-dynamic provisioning.

7.10 THE GFP PROTOCOL

The Generic Framing Procedure (GFP) is a flexible encapsulation framework for traffic adaptation of synchronous broadband transport applications (DS-n/E-n), packetised data (IP, GbE, IP, FC, etc), as well as virtual concatenated NG-S frames with improved bandwidth utilization and efficiency by using LCAS. It supports client control functions that allow different client types to share a channel and it provides an efficient mechanism to map broadband data protocols (such as Fibre Channel, ESCON, FICON, GbE) onto multiple concatenated STS-1 payloads in a revised SONET/SDH frame [48, 49].

The GFP protocol supports mapping of a physical or logical layer signal to a byte synchronous channel, supports different network topologies (short reach, intermediate reach and long reach), exhibits low-latency of packet-oriented or block-coded data streams, and supports differentiated quality of service (QoS) meeting service level agreement (SLA) requirements. Moreover, the GFP allows for existing circuit switching, SONET/SDH, GbE and other packet-based protocols to be used as an integrated and interoperable transport platform that provides cost efficiency, QoS and SLA as required by the client.

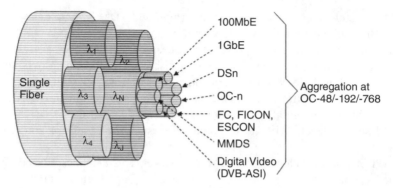

Figure 7.19. **NG-S over WDM** may carry diverse client signals to improve bandwidth utilization and efficiency.

Accordingly, the Next Generation SONET/SDH differentiates from legacy SONET/SDH with its flexible encapsulation of diverse protocols onto GFP generalized frames which are mapped onto synchronous payload envelopes (SPE) of SONET/SDH to support both long packets and circuit switching services.

GFP over the Next Generation SONET/SDH considers that wavelength division multiplexing technology, both coarse and dense (CWDM, DWDM), which is the technology of choice in optical networks. Thus, with the flexibility of GFP over the Next Generation SONET/SDH over WDM a single optical channel (wavelength) may carry diverse client signals to improve bandwidth utilization and efficiency, Figure 7.19.

7.10.1   GFP Header, Error Control and Synchronization

GFP defines different length frames and different client-specific frame types for payload and for management. To increase transmission efficiency, it time multiplexes different frames, frame by frame. If there are no client frames to be multiplexed, it multiplexes idle frames in order to provide a continuous bit stream on the transmission medium.

GFP defines a payload area with its own control fields that include linear point-to-point and ring extensions. Thus, payloads from different clients but of the same type, such as GbE, may be multiplexed using either a sequential round-robin method for synchronous payloads such as DS1 and DS3, or a well-established queuing schedule for asynchronous client payloads with substantial variability in frame length and time of arrival.

GFP defines a flexible frame structure. The GFP frame defines a "core header" and a "payload area". The "core header" supports non-client specific data link management functions such as delineation and payload length indication (PLI) and core header error control (cHEC).

GFP segregates error control between the GFP adaptation process and user data. This allows for sending to the intended receiver frames that have been corrupted under the assumption that end-users will use their own error-correcting codes. In synchronous

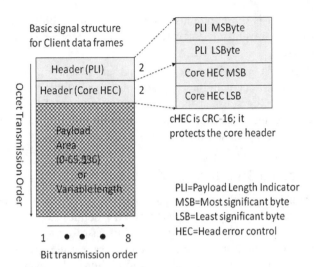

Figure 7.20. The GFP core header.

applications such as video and audio, even corrupted frames are better than no frames at all particularly if they can be restored by the end-user.

GFP defines two classes of functions, *common* and *client-specific.*

- The common is for PDU delineation, data link synchronization, scrambling, client PDU multiplexing and client-independent performance monitoring.
- The client-specific is for mapping client PDU into the GFP payload, and for client specific OA&M (operations, administration, and maintenance).

GFP distinguishes two types of client frames, *client data frame* (CDF), and *client management frames* (CMF). A CMF transports information related to GFP connection or client signal management. A CMF is a powerful feature that allows clients to control the client-to-client connection.

The GFP core header consists of four bytes only, Figure 7.20. The first two define the length of the payload and the other two a standard cyclic redundancy check code (CRC-16) that protects the integrity of the core header from errored bits. The core header error control (cHEC) is able to identify multiple errors and correct single errors.

For frame delineation (or frame synchronization), GFP uses the cHEC; this is a variant of the head-error-control (HEC) and delineation scheme used by ATM [ITU-T Recommendation I.432.1]. Because bit-rate impacts bit error rate (BER), cHEC also plays a significant role in the quality of synchronization.

GFP defines *two scrambling operations*, scrambling the core header and scrambling the payload; scrambling is performed at the transmitter and de-scrambling at the receiver.

- The core header is short and thus is scrambled using a bit by bit complement's-two operation (exclusive OR) with a simple code, such as the pattern 0xB6AB31E0.
- The payload area is much longer and thus a self-synchronizing scrambling algorithm is used. All octets in the payload area are scrambled with the $x^{43}+1$ polynomial. Scrambling starts at the first octet after the cHEC field, it is disabled after the last octet of the frame, and its state is preserved to continue with the following frame, again starting after the cHEC field.

7.10.2   GFP Frame Structure

The type of payload mapped in the GFP payload area is indicated by the binary value of an eight-bit byte, the user payload type (UPI). GFP defines two frame types, the *user data frame* and the *client management frame*. GFP also defines *idle frames* that are used to fill time during the frame multiplexing process. An idle frame consists of four octets, the core header, with all four fields set to all-zero. However, after scrambling, the all-zero code changes to one with sufficient density of ones. Client data frames have priority over management frames, whereas idle frames have the least priority.

7.10.3   GFP-F and GFP-T Modes

GFP specifies two different client types, also known as transport modes, within the same transport channel, the Frame-Mapped GFP (GFP-F) and the Transparent-Mapped GFP (GFP-T).

- The GFP-F mode is optimized for packet switching applications including IP, native point-to-point protocol (PPP), Ethernet (including GbE and 10GbE), and generalized multi-protocol label switching (GMPLS).
- The GFP-T mode is optimized for applications that require bandwidth efficiency and delay-sensitivity applications to include Fibre Channel (FC), FICON, ESCON, and storage area networks (SAN).

To meet real-time requirements, GFP-T considers a fixed frame length and it operates on each character of a frame as it arrives; thus, it does not require buffering and it does not remove idle packets. As a consequence, special MAC is not required. In this case, in order to meet the fixed frame requirement GFP-T defines superblocks. It segments the client signal in eight octets or 64 bits. To each segment an overhead flag bit is added to form a block. Eight consecutive blocks form a superblock; however, all eight flags are collected to form an octet that is placed at the trailing end of the superblock along with 16 CRC bits. The CRC-16 error check code is calculated over all 536 bits (8x8x8+3x8) in the superblock and is added at the end of the superblock.

In summary, according to the general scheme of NG-S, synchronous traffic is mapped in SONET/SDH (VT/TU groups), data (IP, Ethernet, FC, PPP) is encapsulated in GFP (first in GFP client-dependent and then in GFP common to all payloads) and then mapped into the payload envelope of NG-S STS-n.

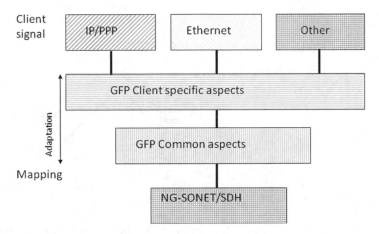

Figure 7.21. Adaptation of various payloads over GFP over NG-S.

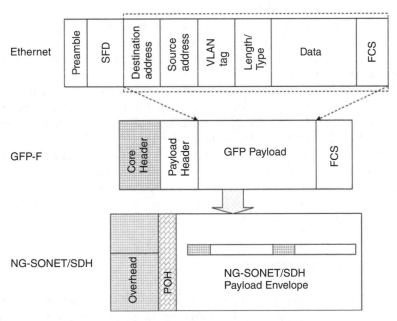

Figure 7.22. Encapsulation of Ethernet frames over GFP and mapping onto the NG-S frame.

During the adaptation to GFP and mapping in NG-S process, overhead bytes and pointers are formed to construct a NG-S frame. Figure 7.21 depicts the logical major steps to map client traffic over NG-S.

An example of encapsulating Ethernet over GFP and mapping onto NG-S frames is shown in Figure 7.22.

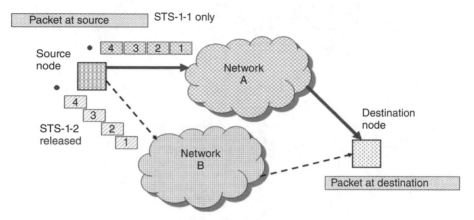

Figure 7.23. LCAS allows for dynamic traffic balancing over NG-S containers.

7.11   THE LCAS PROTOCOL

The *Link Capacity Adjustment Scheme* (LCAS) allows containers in the NG-S to be added or removed dynamically in order to meet user bandwidth requests and to also balance traffic load; addition or deletion of containers should be without traffic flow interruptions (or hitless), Figure 7.23.

LCAS is accomplished using control packets to configure the path between source and destination; it thus adopts a decentralized control process. The control packet is transported over the H4 byte (of SONET/SDH) for high order VC in a superframe, and over the K4 byte (of SONET/SDH) for low order VC [50].

A superframe consists of N multiframes where each multiframe consists of 16 frames. A control packet in the current superframe describes the link status of the next superframe. Changes are proactively sent to the receiving node allowing for ample time to reconfigure itself. Thus, when a data packet arrives, the link reconfiguration has been completed and packet switching takes place without delay.

7.12   THE LAPS PROTOCOL

The *Link Access Procedure-SDH* (LAPS) encompasses data link service and protocol designed to transport point-to-point IP or Ethernet traffic over legacy SONET/SDH.

ITU-T (X.86, pg 9) defines LAPS as "a physical coding sub-layer which provides point-to-point transferring over *SDH virtual containers* and *interface rates*." That is, ITU-T specifies LAPS as a low cost physical coding sub-layer to transport point-to-point IP or Ethernet traffic over SDH virtual containers and interface rates and at the same time provide low latency variance, flow control in bursty traffic, capability of remote-fault indications, ease of use and ease of maintenance. Encapsulation of IPv4, IPv6, PPP and other layer protocols is accomplished with the *service access point identifier* (SAPI).

Starting Flag (0x7E)	Address field (0x04)	Cntl field (0x03)	Payload ID (SAPI)	Data (IPv4/IPv6 or Ethernet)	FCS (CRC)	End Flag

Starting flag = It contains the fixed code 0x7E (0111 1110)
Address field = It contains the fixed 0X04 (0000 01000)
Control field = It contains the fixed code 0x03 (0000 0011)
Payload ID = Two octets define the service access point identifier,
                    or the type of data. For example,
                        0xFE01 identifies Ethernet
                        0x0021 identifies IPv4
                        0x0057 identifies IPv6
Data field = It contains an IP or Ethernet packet
Frame Check Sequence (FCS) field = It contains a 32-bit CRC calculation
                                compliant with RFC 2615
End flag = It contains an octet flag marking the end of the LAPS frame

Figure 7.24. Frame structure of LAPS protocol.

Two ITU-T documents address the encapsulation and rate adaptation of IP and Ethernet over LAPS: ITU-T X.85/Y.1321 defines IP over SDH using LAPS [51] and ITU-T X.86 defines Ethernet over LAPS [52, 53]. ITU-T X.85 & X.86/Y.1321 Recommendations treat SONET/SDH transports as octet-oriented synchronous point-to-point links. Thus, frames are octet-oriented synchronous multiplex mapping structures that specify a series of standard rates, formats and mapping methods. Control signals are not required, and a self-synchronous scrambling/descrambling ($x^{43}+1$) function is applied during insertion/extraction into/from the synchronous payload envelope. The frame structure of LAPS is shown in Figure 7.24.

To facilitate the encapsulation of IP and Ethernet packets over LAPS, an octet staffing procedure is defined (in ITU-T X.85 and X.86) known as *transparency*. Since each frame begins and ends with the same flag (hex 0x7E or binary 0111 1110), there is a probability that 0x7E may be encountered in the information field and thus emulate the frame flag. To avoid this, at the transmitter, an occurrence of 0x7E is converted to the sequence {0x7D 0x5E}. Additionally, occurrences of 0x7D are converted to the sequence {0x7D 0x5D}. The receiver recognizes these sequences (0x7D 0x6E, 0x7D 0x5D) and replaces them with the original octets. Full transparency is also guaranteed (in ITU-T X.86) for Ethernet over LAPS, and LAPS over SDH.

The LAPS mapping of asynchronous data frames over the synchronous SONET/SDH requires *rate adaptation*. That is, asynchronous frames must be buffered and adapted to the SDH rate using "empty" (or idle) payload, which at the receiver it is recognized and removed. In LAPS, the code that fills idle payload is a sequence of {0x7D 0xDD} as long as necessary. Rate adaptation is performed right after transparency processing and before the end flag is added. At the receiver, the rate adaptation sequence of {0x7D 0xDD} is detected and removed in a reverse order: right after the end flag is detected and before transparency.

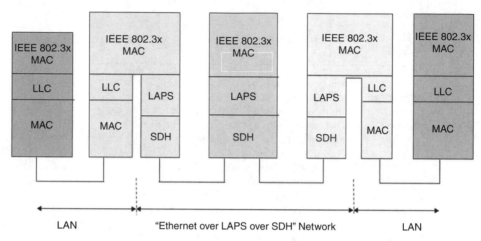

Figure 7.25. Protocol configuration of Ethernet over LAPS over the next generation SONET/SDH.

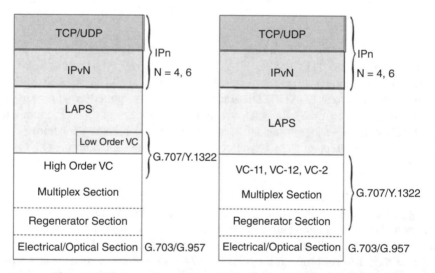

Figure 7.26. Layer/Protocol stack for IP over STM-N using LAPS.

Figure 7.25 shows the protocol configuration of Ethernet over LAPS over the next generation SONET/SDH according to ITU-T Recommendation X.86 and its amendment.

7.13   ANY PROTOCOL OVER SONET/SDH

7.13.1   EXAMPLE 1: IP over LAPS over NG-S

Figure 7.26 illustrates the Layer/Protocol stack for IP over STM-N using LAPS, and for IP over subrate STM-n (sSTM-n) using LAPS over SDH according to ITU-T Rec.

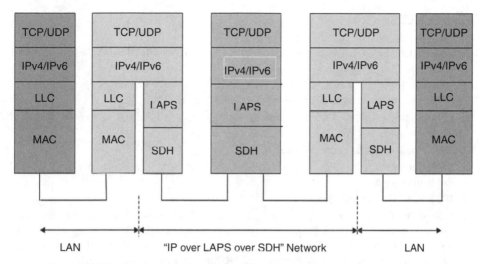

Figure 7.27. Protocol configuration of IP over LAPS over legacy SONET/SDH.

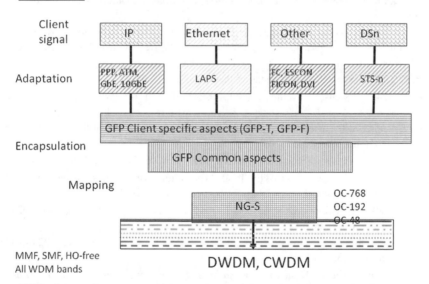

Figure 7.28. Adaptation, encapsulation and mapping over NG-S over WDM for a variety of payloads.

X.85/Y.1321, whereas Figure 7.27 illustrates the protocol configuration of IP over LAPS over legacy SONET/SDH.

7.13.2 EXAMPLE 2: Any Payload over LAPS over Next NG-S over WDM

Figure 7.28 illustrates an example for a variety of payloads undergoing adaptation, encapsulation and mapping over NG-S over WDM.

REFERENCES

1. S.V. Kartalopoulos, *Understanding SONET/SDH and ATM*, IEEE-Press/Wiley, Piscataway, NJ, 1999.
2. Telcordia (previously Bellcore), GR-253-CORE, Issue 2, "Synchronous Optical Network (SONET) Transport Systems, Common Generic Criteria", 1992.
3. J. van Bogaert, "E-MAN: Ethernet-based metropolitan access networks", *Alcatel Telecommunications Rev.*, 1ˢᵗ Quarter, 2002, pp. 31–34.
4. IEEE 802.3ab, standard on 1000BaseT.
5. D.G. Cunningham, and W.G. Lane, *"Gigabit Ethernet Networking"*, MacMillan Technical Publishing, 1999.
6. IETF RFC 791, Internet protocol.
7. "'Super Wi-Fi' nears final approval in U.S.". CBC. 13 September 2010. http://www.cbc.ca/technology/story/2010/09/13/super-wifi-white-spaces-fcc.htmlREF1. "Securing Wi-Fi Wireless Networks with Today's Technologies". Wi-Fi Alliance. 2003-02-06. http://www.wi-fi.org/files/wp_4_Securing_Wireless_Networks_2-6-03.pdf. Retrieved Nov 23, 2010.
8. S.V. Kartalopoulos, *Security of Information and Communication Networks*, IEEE/Wiley, 2009.
9. "WPA™ Deployment Guidelines for Public Access Wi-Fi® Networks". Wi-Fi Alliance. 2004-10-28. http://www.wi-fi.org/files/wp_6_WPA_Deployment_for_Public_Access_10-28-04.pdf. Retrieved Nov 23, 2010.
10. RFC 3748, "Extensible Authentication Protocol (EAP)", June 2004.
11. RFC 5247, "Extensible Authentication Protocol (EAP) Key Management Framework", August 2008.
12. FCC Sec.15.249, "Operation within the bands 902–928 MHz, 2400–2483.5 MHz, 5725–5875 MHZ, and 24.0–24.25 GHz".
13. "WiMax Forum", http://www.wimaxforum.org/, Retrieved Nov. 23, 2010.
14. "Facts About WiMAX And Why Is It 'The Future of Wireless Broadband'", http://www.techpluto.com/wimax-in-detail/, Retrieved Nov. 23, 2010.
15. "IEEE 802.16 WirelessMAN Standard: Myths and Facts", http://www.ieee802.org/16/docs/06/C80216-06_007r1.pdf, Retrieved Nov. 23, 2010.
16. "IEEE 802.16e Task Group (Mobile WirelessMAN)", http://www.ieee802.org/16/tge/, Retrieved Nov 23, 2010.
17. K. Fazel and S. Kaiser, *Multi-Carrier and Spread Spectrum Systems: From OFDM and MC-CDMA to LTE and WiMAX*, 2nd Ed., John Wiley & Sons, 2008.
18. M. Ergen, *Mobile Broadband—Including WiMAX and LTE*, Springer, NY, 2009.
19. ITU, "What really is a Third Generation (3G) Mobile Technology", (PDF), http://www.itu.int/ITU-D/imt-2000/DocumentsIMT2000/What_really_3G.pdf. Retrieved Nov 23, Retrieved Nov 23, 2010.
20. ITU-R. RP-090743 TR TR36.912 v9.0.0, and also ANNEX A3, C1, C2 and C3.
21. LTE Encyclopedia: http://sites.google.com/site/lteencyclopedia/ accessed March 2011.
22. White Paper: LTE Protocol Overview http://www.scribd.com/doc/18094043/LTE-Protocol-Overview; accessed March 2011.
23. 4G LTE Advanced Tutorial http://www.radio-electronics.com/info/cellulartelecomms/lte-long-term-evolution/3gpp-4g-imt-lte-advanced-tutorial.php accessed March 2011.

24. Telcordia (previously Bellcore), GR-253-CORE, Issue 2, "Synchronous Optical Network (SONET) Transport Systems, Common Generic Criteria", 1992.

25. American National Standard for Telecommunications—Synchronous Optical Network (SONET): Physical Interface Specifications", ANSI T1.106.06, 2000.

26. ANSI T1.102-1993, *Telecommunications–Digital hierarchy- Electrical interfaces*, 1993.

27. ANSI T1.107-1988, *Telecommunications–Digital hierarchy-Formats specifications*, 1988.

28. ANSI T1.105.01-1994, *Telecommunications–Synchronous optical network (SONET)-automatic protection switching*, 1994.

29. ANSI T1.105.03-1994, *Telecommunications–Synchronous optical network (SONET) Jitter at a network interfaces*, 1994.

30. ANSI T1.105.04-1994, *Telecommunications–Synchronous optical network (SONET)-data communication channel protocols and architectures*, 1994.

31. ANSI T1.105.05-1994, *Telecommunications–Synchronous optical network (SONET)-Tandem connection maintenance*, 1994.

32. IETF RFC 2823, PPP over Simple Data Link (SDL) using SONET/SDH with ATM-like framing, May 2000.

33. ITU-T Recommendation G.707/Y1322, "Network Node Interface for the Synchronous Digital Hierarchy (SDH)", Oct. 2000

34. ITU-T Recommendation G.783, "Characteristics of Synchronous Digital Hierarchy (SDH) Equipment functional blocks", Feb. 2001

35. ITU-T Recommendation G.784, "Synchronous Digital Hierarchy (SDH) management", 1998

36. ETSI European Standard EN 300 417-9-1 (currently ITU-T), "Transmission and Multiplexing: Generic requirements of transport functionality of equipment; Part 9: Synchronous Digital Hierarchy (SDH) concatenated path layer functions; Sub-part 1: Requirements".

37. ITU-T Recommendation G.828, *"Error performance parameters and objectives for international, constant bit rate synchronous digital paths"*, Feb. 2000.

38. Telcordia (previously Bellcore), TA-NWT-1042, "Ring Information Model", 1992.

39. S.V. Kartalopoulos, *Next Generation SONET/SDH: Voice and Data*, Wiley/IEEE Press, 2004.

40. S.V. Kartalopoulos, *Introduction to DWDM Technology: Data in a Rainbow*, Wiley/IEEE Press, 2000.

41. S.V. Kartalopoulos, *Fault Detectability in DWDM*, IEEE Press/Wiley, 2002.

42. S.V. Kartalopoulos, *Next Generation Intelligent Optical Networks*: From Access to Backbone, Springer, 2008.

43. H. Christiansen, T. Fielde, and H. Wessing, "Novel label processing schemes for MPLS", *Opt. Networks Mag.*, vol. 3, no. 6, Nov/Dec 2002, pp. 63–69.

44. E. Rosen et al., *Multiprotocol Label Switching Architecture*, IETF RFC 3031, Jan. 2001.

45. Y-D. Lin, N-B. Hsu, R-H. Wang, "QoS Routing Granularity in MPLS Networks", *IEEE Communications Mag.*, vol. 40, no. 6, June 2002, pp. 58–65.

46. P. Ashwood-Smith, et al., "Generalized MPLS—CR-LDP Extensions", IETF RFC 3472, January 2003.

47. L. Berger, et al., "Generalized MPLS—RSVP-TE Extensions", IETF RFC 3473, January 2002.

48. ITU-T recommendation G.7041/Y.1303, "The Generic Framing Procedure (GFP) Framed and Transparent", Dec 2001.

49. E. Hernandez-Valencia, "Generic framing procedure (GFP): A next-generation transport protocol for high-speed data networks", *Opt. Networks Mag.*, vol. 4, no. 1, Jan/Feb 2003, pp. 59–69.

50. ITU-T Recommendation G.7042/Y.1305, "Link Capacity Adjustment Scheme (LCAS) for Virtual Concatenated Signals", Nov. 2001.

51. ITU-T Recommendation X.85/Y.1321, "IP over SDH using LAPS", 03/2001.

52. ITU-T Recommendation X.86, "Ethernet over LAPS", 02/2001.

53. ITU-T Recommendation X.86/Y.1323, Amendment 1, "Ethernet over LAPS, Amendment 1: Using Ethernet **flow** control as rate limiting", 04/2002.

8

FSO NETWORK SECURITY

8.1 INTRODUCTION

The integrity of legacy copper-based networks is known to have been compromised by eavesdropping using simple electrical means. Accessing the loop side of a circuit-switched communications network required moderate networking know-how in order to tap a 2-wire pair and eavesdrop a conversation, but advanced know-how to mimic signaling codes, with the so called "blue box", in order to establish end-to-end connectivity avoiding charges. However, the latter intervention was identified and enhancements in the network signaling protocol and signaling method eliminated the use of the illegal "blue-box". Similarly, at the core network, demultiplexing end-user time-slots required specialized equipment so that the network was not challenged with eavesdropping and virus attacks by outside bad actors. Virus and other malicious soft attacks appeared with the spread of the software-based Internet nodes and "cyber-security" was born to deter attempts on the network and end terminals, PCs and the like; in fact, "cyber-security" and "software virus" are terms that go hand-in-hand with Internet and computer communication networks [1].

The differences between the synchronous public switched digital network (PSDN) and the data networks (such as the Internet) are:

Free Space Optical Networks for Ultra-Broad Band Services, First Edition. Stamatios V. Kartalopoulos.
© 2011 Institute of Electrical and Electronics Engineers. Published 2011 by John Wiley & Sons, Inc.

- the latter transports packets in an asynchronous manner in contrast to
- the former that transports byte-size information in a continuous and synchronized manner (such as digitized voice samples every 125 μsec, or real-time video) constraints with strict real-time requirements.

When packets enter a data node or router, they are buffered in a queue until they are switched to the output buffer. In general, the route of a "connectionless" network is not under network control and it is determined with one of several methods, depending on network protocol and quality of data service. It is the data buffering in routers that allow smart but malicious programs to creep into their computer-based execution ability and initiate one of many undesirable actions such as spoofing, cloning, file deletion, file copying, data harvesting, and so on.

In contrast, the public switched digital network (PSDN) is based on standardized synchronized frames, such as for example DS1 and SONET/SDH [2], which flow through a switching or cross-connecting node since the path has been established during the call initiation procedure. Thus, end-user data in routers can be remotely accessed with specialized know-how tools whereas end-user data is transparent to circuit-switched networks.

Besides the PSDN and the data packet network (such as the Internet), the cellular wireless network is also vulnerable to eavesdropping and to calling number mimicking. In fact, accessing calling numbers and pin codes from the airwaves has been a relatively easy task. Therefore, in order to reduce risk, enhanced user authentication and secure calling number identity procedures have been included in the cellular wireless protocols.

The glassy fiber in optical networks transports Gb/s per optical channel or Tb/s per fiber. One Tb/s corresponds to more than 10 million simultaneous conversations, or about half a million simultaneous video channels, or many millions of documents and transactions, many of which may be classified [3].

Although optical technology is more complex than its predecessors, it attracts bad actors because of the huge amount of information that transports. Bad actors with the proper know-how and sophisticated tools may attack the medium and harvest huge amounts of information, mimic the source, alter information or disable the proper operation of the network. Thus, in order to assure data security and integrity as well as privacy, highly complex, difficult and sophisticated algorithms have been developed. However, in addition to this the network itself should be sophisticated and able to detect malicious attacks and to outsmart them by adopting sophisticated countermeasure strategies.

In general, information assurance, integrity and security is a concern that aims to ensure a level of trust commensurate with client expectations. Expectations include information or data protection at the application layer, at the layer boundaries of the reference model (such as the ISO, ATM, TCP/IP), at the transport layer and in the computation environment. Information security also aims to ensure that the access to information and to the network is controllable and capable for self-defense and able to initiate countermeasures.

8.2 CRYPTOGRAPHY

Cryptography is a process that transforms data (or *plain text*) into unintelligible and undetectable data, *the cipher text*, to all but the rightful recipient; the latter is the only one that possesses special knowledge and proper authorization to transform the cipher text back to its original form.

Unintelligible means that if the cipher text is viewed by a third party, an eavesdropper, the original text cannot be extracted from it. Similarly, undetectable means that if an unauthorized party looks for a particular cipher text, the cipher text cannot be singled out.

Cryptography performs several important information security services: source and destination *authentication*, *authorization*, data *integrity*, message *confidentiality*, and *non-repudiation* [4].

The cryptographic process includes a secure mechanism for transporting the special knowledge to the rightful recipient. It also includes a process for transforming the cipher text back to its original form. The cryptographic process that changes the form of the original data (the *plain text*) to unintelligible is called *encoding* or *ciphering*. The result of encoding is the *cipher text*. The process of restoring the original form of data is called *decoding* or *deciphering*.

In order to produce a cipher text that only the rightful recipient can read, the sender needs a secret code known as *cipher key*. The method by which the cipher key is exchanged and agreed upon by both sender and receiver is a process known as *key establishment*. When the key has been agreed upon, it may be registered in a *key depository* or *key management archive* by a key *registration authority*. Similarly, a *public key certificate* uniquely identifies an entity, it contains the entity's public key, it uniquely binds the public key with the entity, and it is digitally signed by a trusted *certificate authority*, as part of the X.509 protocol (or ISO Authentication framework). If the key is suspected to be *compromised* it may be *revoked*. In this case, the key establishment may restart and the key may be *updated*.

The method of encoding/decoding with the same key is known as *symmetric cryptography* and the key is known as a *symmetric key*. To generate the symmetric key another symmetric key is needed first, the *seed key*, which is used to encode the final key, hence known as the *key encrypting key*; this process is known as *key wrapping*.

In certain applications where the text is binary serial, the cipher key operates on the serial bit-stream bit or byte at a time using the XOR function. Such cryptographic methods are also known as *stream ciphers* denoting the serialization of data. The stream cipher text is decoded serially by applying the same XOR logic operation with the same key. The key in this case is generated by a pseudorandom generator, which in this case it is also called a *key-stream generator*.

One of the issues with the symmetric cryptography method is the *key distribution*. That is, the transportation or communication of the key from the sender of the cipher-text to the rightful recipient of it. If the cipher-key is intercepted by an intruder during its transportation, the value of the cipher-text is meaningless in cryptography. However, if the secrecy of key distribution is an issue, symmetric cryptography is very strong since the key can be as random as possible and as long as needed.

If the cryptographic method uses one key for encoding and another for decoding, then the cryptographic method is known as *asymmetric cryptography*. This method is particularly interested because intercepting the encoding key during its transportation does not reveal the decoding key. Thus, the asymmetric key method is appealing and it is the method that has inspired some complex cryptographic methods applicable to the Internet and other applications. Among them is the *public key cryptography*.

Public key cryptography uses a pair of two key ciphers, one which is public and another which is private. Plain-text encrypted with the public key can only be decrypted with the associated private key. The public key cryptography is also used in conjunction with one-way hash functions to produce a digital signature. According to this, messages signed with the private key can be verified with the public key [5–7].

8.3   SECURITY LEVELS

A cryptographic module typically contains encryption subunits, cryptographic public and private keys, plaintext or cipher text, memory, and registers. A module has data input-output ports, a maintenance port, a door or a cover that potentially may be physically accessed, or be affected by electromagnetic interferences, environmental fluctuation, or power fluctuation. Thus, a bad actor may exploit the weaknesses and vulnerabilities of the module to gain unauthorized access into the cryptographic keys. Security management provides requirements that safeguard the physical entity of the module.

FIPS 1402 defines four security levels [8–12]. A synopsis of the four levels follows:

- Security Level 1 (SL-1) specifies basic requirements for a cryptographic module, such a personal computer encryption board, so that software and firmware components are executed on an unevaluated general purpose computing environment. No specific physical requirements are provided for a SL-1 module. When performing physical maintenance, all plaintext secret and private keys and other *critical security parameters* (CSP) contained in the cryptographic module shall be zeroed; this is performed procedurally by the operator or autonomously by the cryptographic module
- Security Level 2 (SL-2) protects against unauthorized physical access by adding the requirement for tamper-evidence coatings, seals, and locks on door and covers. At this level, when an operator needs to gain access to perform a corresponding set of services, the operator requests authorized access and the cryptographic module authenticates the authorization access. To accomplish this, the SL-2 requires at minimum role-based authentication. SL-2 enhances LS-1.
- Security Level 3 (SL-3) attempts to prevent an intruder from gaining access to CSPs in the cryptographic module. Thus, SL-3 requires physical security mechanisms that exhibit high probability of detecting attempts to physical access as

well as reporting mechanisms. When the enclosure is violated it should cause serious damage to the cryptographic module. To accomplish this, SL-3 requires identity-based authentication mechanisms. SL-3 requires physically separate input-output ports for ciphered text or plaintext to be ciphered by the cryptographic module. Circuitry detects an attempt to open the cover or the door of a module or to access the maintenance access interface. Then, the circuitry zeroes the plaintext secret and public cryptographic keys and CSPs contained in the cryptographic module. SL-3 enhances LS-2.

- Security Level 4 (SL-4) provides a complete envelope of physical security around the cryptographic module and it detects any attempt of penetrating the cryptographic module to gain access of the cryptographic keys. SL-4 also protects against security compromise as a result of external environmental avert conditions and fluctuations. The cryptographic module contains either *environmental failure protection* (EFP) or *environmental failure testing* (EFT) mechanisms. Attempting to remove or dissolve the protective coating should result in serious damage of the cryptographic module.

8.4 SECURITY LAYERS

In modern communication networks, security is not limited to ciphering user information (data) messages only. Data enters computer-based nodes and they are temporarily stored in memory. In addition, nodes are provisioned and are maintained with network control messages. Messages are transmitted electronically, optically or over radio waves, which potentially may be intercepted.

For example, encrypted data implies that the transport medium is secure. This may be correct if we had an encryption algorithm that generates and distributes keys that cannot be broken by any means. So, encrypting data is only one of the dimensions in communications security, which is the responsibility of end users. We call this *security at the information layer*.

Assume that a bad actor taps the physical layer of the communications network (the link or the node) to eavesdrop and copy data, even if data is encrypted. If the bad actor could extract the cipher key, then he/she could decipher the encrypted text, and perhaps alter it, encode it again and retransmit it to its destination. In this case, monitoring and detecting network malicious interventions require intelligent methods. Thus, *network physical layer security* is another dimension, which is the responsibility of network provider.

Assume that a bad actor is not interested in eavesdropping messages but in disabling the network from establishing the cipher key, or in destroying messages in routers and end-terminals, or harvesting data from computer based nodes, or altering the security and destination address, or altering the data field. This could be accomplished on-site or remotely, and it could cause network congestion, node or router reconfiguration, or to plant in it an executable program that can be activated under specific conditions (a command or a specific time), disable the authentication and the key distribution process, and so on. This is another dimension of *security on the MAC/*

Network layer. Thus, monitoring and detecting proper network protocol execution is another dimension in security.

In addition, the network should include intelligent mechanisms to outsmart malicious actors and to avoid malicious events; such mechanisms are known as *countermeasures*.

8.4.1 Security on the Information Layer

Consider a classified text to be transmitted. Then, some steps should be taken to make it unintelligible to a third unauthorized party. The first level of assurance is to encrypt the text. Ciphering text at the source and deciphering at the receiving end constitutes the Information security layer. This layer is not concerned with the transporting mechanism itself but with:

- Which algorithm at the source can encrypt the message and which one can decrypt it at the rightful destination, so that even if it is intercepted by a third party it remains unbreakable?
- How can a secure key be established between the source and the destination before the encrypted message is transported?

In general, cryptographic keys are classified as symmetric and as asymmetric. Typically, the security level that is offered by the cryptographic method depends on the difficulty a third party to capture and extract the cipher key. Similarly, the efficiency of the cryptographic method depends on the speed the cipher text is converted to plain text and also on the length of the key; the shorter the key, the faster the deciphering but also the easier to compute the key by a third party.

In this chapter, we examine certain cryptographic methods, some simple (as already described) and some difficult; in fact, the most difficult ones depend on the difficulty to compute algebraic algorithms and also on special properties of quantum particles.

8.4.2 Security on the MAC/Network Layer

The communications media access control layer (MAC) is computational-based and is responsible for reliable data delivery, network access and security (user authentication, authorization). In general, the MAC layer grants access to requests to transmit, calculates the best path/route in the network, determines the integrity of data frames, discards frames, retransmits frames, or reroutes frames. Thus, this layer is critical to the proper operation of the network and if it does not function properly, it may cause node congestion or even network congestion. Therefore, the security of this layer is also important. The security of a node is assured by proper screening (firewalls) and by dynamically updated access passwords. However, the data network has proven repeatedly to be vulnerable to attacks as a result of packet *store and forward* or *partially store and forward* in routers. Thus, a malicious executable program that hides in a packet may enter the router at the MAC layer. Virus planting, cloning, spoofing, flooding, Trojan horses and so on are among the malicious attacks of bad actors who do not even have to be in proximity to the node under attack.

8.4.3   Security on the Link Layer

The link layer encompasses the communications link from the transmitter to the receiver, including the medium. In communication networks, the medium may be guided or unguided. Among the guided is twisted pair copper (TP), coaxial cable (CB) and single mode fiber (SMF). Among the unguided are electro-magnetic waves in atmosphere or in space, or a free space optical laser beam in atmosphere.

The link length may be from hundreds of meters to many kilometers. Thus, it is possible that the link also includes repeaters, add-drop multiplexers, and switches or routers. As a result, even if we can trust the physical integrity of modules on the link, the medium itself offers an opportunity for attacks and it cannot be trusted; twisted copper pairs and coaxial cables have been easily tampered. The fiber medium, although difficult, it is not impossible to tamper with proper equipment. In unguided media, the electromagnetic waves in the atmosphere reach both friendly and foe antennas and therefore this medium is the most vulnerable to eavesdropping and to source mimicking. The free space optical beam is relatively secure because the beam is very narrow, it is invisible to the human eye and it requires line of sight to operate, although there are vulnerabilities, which we explore separately.

The transmission medium, besides transporting information (ciphered text), it also transports the cipher key during the key distribution process. Thus, if a bad actor is able to capture or to compute the key in reasonable time, then the ciphertext can be compromised.

Therefore, the communications link should be able to continuously monitor its integrity, detect interventions, authenticate the channel and link identification or signature, and include countermeasure strategies.

8.4.4   FSO-WLAN Security

When FSO backhaul is integrated with WLAN technology and protocols, in addition to security mechanisms described previously, the wireless network is on a critical path and its security mechanism need to be evaluated.

- If the Wi-Fi technology and protocol is used, the Wi-Fi Alliance has put emphasis on the Wi-Fi Protected Access (WPA) security mechanism. WPA utilizes the Temporal Key Integrity Protocol (TKIP), which is part of the IEEE 802.11i standard, to provide data encryption as well as user authentication based on the Extensible Authentication Protocol (EAP), and IEEE 802.1X, which is also a mechanism for port-based network access control [13]. The EAP, defined by IETF (RFC 3748), is a flexible framework that allows authentication protocols to be exchanged between the end user and the authenticator.
- If the WiMAX technology and protocol is used, WiMAX security supports encryption standards and end-to-end authentication protocols. All traffic on

WiMAX is encrypted using the Counter Mode with Cipher Block Chaining Message Authentication Code Protocol (CCMP), which uses the Extensible Authentication Protocol (AES) for transmission security and for data integrity authentication. Because WiMAX is subject to IP vulnerabilities, such as denial of service (DOS) and attacks by malicious hackers, the continuous intrusion detection is important.

Between the mobile station (MS) and the base station (BS), EAP utilizes the protocol defined in IEEE 802.16e-2005 and it runs over the WiMAX physical layer (PHY) and the media access control (MAC) layer.

The mobile WiMAX end-to-end network architecture model follows the network reference model (NRM) developed by the WiMAX Forum's Network Working Group (NWG) of the WiMAX Forum [14]. Further work on the WiMAX integrated with FSO is still going on [15–18].

8.5   FSO INHERENT SECURITY FEATURES

Advanced optical communication networks require security on all levels, physical, control and application [19].

In general, FSO transceivers are installed on rooftops of tall building and thus the transceivers are not easily accessible, the electronics may be enclosed in secure enclosures and the interconnecting cables may be protected by non-tamper conduits [20].

Additionally, enclosures may have sensors that detect unauthorized intrusions and trigger alarms if tampering is detected.

Finally, because FSO networks are targeted for private network applications, a private encryption protocol with individualized cipher keys may be used over each link.

FSO networks have also several inherent security features:

- the laser beam is very narrow and is in the spectrum that is invisible to human eye; as a result, the invisible FSO link is difficult to localize by a third party.
- the laser beam is at a height (as measured from the ground) that is not easily accessible.

8.5.1   FSO Beam Overspill

Modern eavesdroppers should not be underestimated; (s)he has available the most sophisticated equipment and intelligence. Our eavesdropper, Evan, knows that the beam is slightly divergent forming an acute cone, the diameter of which is proportionally increasing with distance. As such, at the receiver the diameter of the cone is much larger than the aperture of the receiver and some optical power of the signal is transmitted beyond the receiver, known as overspill, Figure 8.1. Overspill presents an opportunity for eavesdropping and it may be exploited by Evan. Therefore, the narrower the beam the less the overspill is.

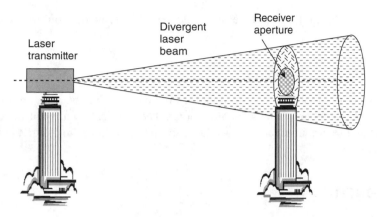

Figure 8.1. The divergent beam overspills the receiver aperture providing an opportunity for eavesdropping.

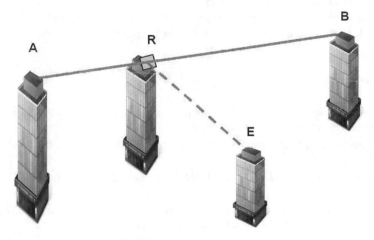

Figure 8.2. A beam from A to B passes near an intermediate building (R), where it may be partially deflected to another building (E) for eavesdropping.

In particular, mesh-FSO networks [21] have more than one transceiver per node and therefore more opportunities for overspill tampering.

8.5.2 Beam Tapping

Although the FSO beam is narrow and invisible, there are methods by which a malevolent eavesdropper can identify the path of the beam. In some cases, the beam may pass nearby another building between the transmitter and the receiver; if the distance from the third building is few meters, then, it is possible to tap the beam by deflecting part of its power with a transparent plate, Figure 8.2.

However, beam tapping is easily detected and remedied by periodic visual inspection.

8.5.3 FSO Cable Tapping

Cable tapping is a typical case. The signal from the transceiver housing on the rooftop is connected via cables to the switch or router. The physical security of the cables is important and secure installation procedures should be followed to assure the integrity of the cables.

8.6 CONCLUSION

In conclusion, security of any communication network is vulnerable to eavesdropping. Malevolent eavesdroppers are sophisticated, educated and specialized in the field, and they have in their disposal sophisticated tools, some of them specially made for it. Therefore, the network security engineer should always be alert, should examine all possible vulnerabilities, develop necessary countermeasures, and continuously monitor the network at its different layers and levels for possible attacks.

REFERENCES

1. S.V. Kartalopoulos, "A Primer on Cryptography in Communications", *IEEE Communications Magazine*, vol. 44, no. 4, pp. 146–151, 2006.
2. S.V. Kartalopoulos, *Understanding SONET/SDH and ATM*, IEEE Press, 1999; also, Prentice Hall of India.
3. S.V. Kartalopoulos, *Introduction to DWDM Technology: Data in a Rainbow*, Wiley/IEEE Press, 2000.
4. S.V. Kartalopoulos, "*Security of Information and Communication Networks*", Wiley/IEEE, 2009; recipient of the "2009 Choice Award of Outstanding Academic Titles".
5. FIPS Pub 186-2, *Digital Signature Standard*, January 2000
6. FIPS Pub 186-2 change notice, *Digital Signature Standard*, October 2001.
7. FIPS 180-2, *Secure Hash Standard (SHS)*, August 2002.
8. FIPS PUB 140-2, *Security Requirements for Cryptographic Modules*, 2002.
9. FIPS PUB 140-2, *Security Requirements for Cryptographic Modules*, Annex A: Approved Security Functions, Draft, 2005.
10. FIPS PUB 140-2, *Security Requirements for Cryptographic Modules*, Annex B: Approved Protection Profiles, Draft, 2004.
11. FIPS PUB 140-2, *Security Requirements for Cryptographic Modules*, Annex C: Approved Random Number Generators, Draft, 2005.
12. FIPS PUB 140-2, *Security Requirements for Cryptographic Modules*, Annex D: Approved Key Establishment Techniques, Draft, 2005.
13. Wi-Fi Alliance "WPA™ Deployment Guidelines for Public Access Wi-Fi® Networks", October 28, 2004.

14. WiMAX Forum, "WiMAX End-to-End Network Systems Architecture - 3GPP/ WiMAX Interworking, Release 1", 2006.

15. Wen Gu, S.V. Kartalopoulos, and P. Verma, "Performance Evaluation of EAP-based Authentication for Proposed Integrated Mobile WiMAX and FSO Access Networks", to be presented at the IEEE Wireless Communications and Networking Conference 2011 (IEEE WCNC 2011 – Network, March 29–31, Cancun, Mexico.

16. Wen Gu, S.V. Kartalopoulos, and P. Verma, "Secure and Efficient Handover Schemes for WiMAX over EPON networks," presented at the 4th WSEAS Conference, January 2010, Harvard University, Cambridge, Mass.

17. Di Jin, S.V. Kartalopoulos, and P. Verma, Chapter 5, *Wireless Ad Hoc and Sensor Networks Security*, in "Security and Privacy in Mobile and Wireless Networking", S. Gritzalis, T. Karygiannis and Ch. Skianis (editors), Troubador, ISBN: 978-1905886-906, 2009.

18. Wen Gu, S.V. Kartalopoulos, and P. Verma, Chapter 7, *Security Architectures and Protocols in WLANs and B3G4G Mobile Networks*, in "Security and Privacy in Mobile and Wireless Networking", S. Gritzalis, T. Karygiannis and Ch. Skianis (editors), Troubador, ISBN: 978-1905886-906, 2009.

19. S.V. Kartalopoulos, "Security in Advanced Optical Communication Networks", Proceedings of the IEEE ICC 2009 Conference, Dresden, GE. June 14–18, 2009, IEEE Catalog no.: CFP09ICC-USB, ISBN: 978-1-4244-3435-0.

20. S.V. Kartalopoulos, "Protection Strategies and Fault Avoidance in Free Space Optical Mesh Networks", IEEE ICCSC'08 Conference, Shanghai, May 26–28, 2008; Proceedings on CD-ROM: ISBN 978-1-424-1708-7.

21. S.V. Kartalopoulos, "Security of reconfigurable FSO Mesh Networks and Application to Disaster Areas", SPIE Defense and Security Conference, March 16–20, 2008, Orlando, Florida, paper no. 6975–9, Session S2; Proceedings on CD-ROM.

9

FSO SPECIFIC APPLICATIONS

9.1 INTRODUCTION

Optical communication has been the preferred method for very-high bandwidth transport of data; currently, communication services demand needs integrated voice, high-speed data, internet, image, video, games, radio and more. This integration has been easily met by the core network using optical technology [1]. However, it has challenged the current access network, which has to interface a variety of data rates, inhomogeneous traffic types and technologies, including wireless. A technology that is able to meet this challenge in the access space is optical and the next generation standards such as Gigabit Ethernet and next generation SONET/SDH [2].

FSO technology has also been applied to communications, as already described, and because of the easy and quick installation to some other (more specific) applications.

In this chapter, the objective is to briefly describe a couple of possible applications and not to provide an exhaustive list; more applications are already in use and many more have already been proposed.

Free Space Optical Networks for Ultra-Broad Band Services, First Edition. Stamatios V. Kartalopoulos.
© 2011 Institute of Electrical and Electronics Engineers. Published 2011 by John Wiley & Sons, Inc.

9.2 FSO NETWORKS FOR HIGHWAY ASSISTED COMMUNICATIONS

Consider a free space optical network in a multi-hop topology able to transfer broadband traffic in excess of 1 Gb/s over many kilometers. The distance of each hop is from 1–4 km, and the overall path length may be from 1–20 km, depending on the number of intermediate nodes between the source and destination.

Intermediate nodes may be optical add-drop multiplexers or pass-through nodes. Straight pass-through nodes are powered by solar cells. In this topology, the path is typically full-duplex, and depending on application, the path transports a balanced or unbalanced traffic.

The aforementioned FSO network may be deployed by a highway to carry data from/to a control center to/from toll booths, highway signs, highway surveillance cameras, and emergency stations. In areas where fog and other atmospheric effects may cause temporary disruption of FSO links, RF back-up may also be used.

9.3 MESH-FSO IN DISASTER AREAS

Free-Space Optical networks are easily deployable and thus they are a viable solution to emergency ad-hoc networks in ring or mesh topology covering several square kilometers and supporting wireless voice, data and video services [3]. Such networks are particularly useful in natural disaster situations, in military expeditions, and in other short term communication network applications. It has been proven that the ring and the mesh network topology has superior disaster avoidance and better protection capabilities. We describe the applicability of FSO in the case of natural disaster.

Natural disasters (floods, hurricanes, storms, eruptions, etc.) that take place in inhabited areas greatly affect the communications infrastructure and particularly the connectivity to the home/business, known as "access" or "last/first mile", affecting broadband and baseband communication, such as telephony, Internet, and video services for quite some time. When a disaster occurs, it is important that an interim communication solution is employed until service is restored; a candidate solution is a mesh-FSO network, preferably with RF back-up.

The FSO technology is used in the optical backhaul network and FSO nodes are integrated with fiber as well as wireless access technologies, such as WiMax (Worldwide Interoperability for Microwave Access) and WiFi.

WiMax is a wireless technology for indoors and outdoors communications and it is based on the IEEE 802.16m standard (an updated IEEE 802.16) and it offers up to 1 Gb/s broadband services [4–11]. WiMax gateways provide WiFi access to multiple devices, Ethernet accessibility, and PSTN (Public Switched Telephone Network) accessibility.

Similarly, the Wi-Fi wireless local area technology, defined by the Wi-Fi Alliance, is based on the IEEE 802.11 standard (802.11n standard is an extension of the original 802.11 to improve the quality of the wireless link and thereby to increase both data rate and link range) [12–14].

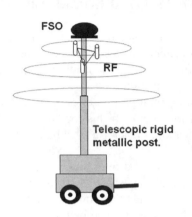

Services
Wired telephony
Mobile (wireless) telephony
Police band
High speed data (Ethernet, DSL, other)
WiMax/WiFi for wireless data devices
FSO links

Equipment
Access system/aggregator
w/switching capability
Ethernet router
Wireless station
WiMax Gateway
WiFi router
DSL ports

Figure 9.1. FSO nodes "on-wheels" are pre-engineered to support both optical and RF connectivity.

Both, the WiMax and Wi-Fi technologies, can be integrated in FSO nodes to quickly construct a mesh topology network, and to establish communication services bypassing the damaged infrastructure. The mesh-FSO network can also be deployed in areas where no previous infrastructure exists in order to meet urgent and temporary communications needs.

To respond quickly to areas affected by a natural disaster, after a quick survey, pre-engineered nodes can be easily moved and positioned to appropriate locations; for ease of deployment, pre-engineered nodes may be "on wheels" (in extreme cases, if necessary, they can be hoisted and positioned by helicopters, Figure 9.1.

Thus, "nodes on wheels" quickly construct a stationary ad-hoc reconfigurable mesh network so that, when connected with the public network (one of the nodes is connected with a central office), it establishes communication services (voice, data, video) from any point of the covered area to anywhere in the world, Figure 9.2. Notice that the high bandwidth of the mesh-FSO network allows for multi-play services (land and mobile voice, Internet, picture and video) necessary to various emergency units such as medical, surveillance, operations, search, police and fire, and more. Then, as soon as the communications infrastructure has been restored, the mesh-FSO can be easily packed and moved to another location, as needed.

9.4   VISUAL LIGHT COMMUNICATION

Visual Light Communications (VLC) is a recent FSO technology that capitalizes on low cost opto-electronic devices, light emitting diodes (LED) and PIN photodetectors operating in the visible spectrum, from 400 nm (deep blue) to 700 nm (deep red); LEDs have already been used in laser-pointers, in flat television panels, and other mass-produced applications.

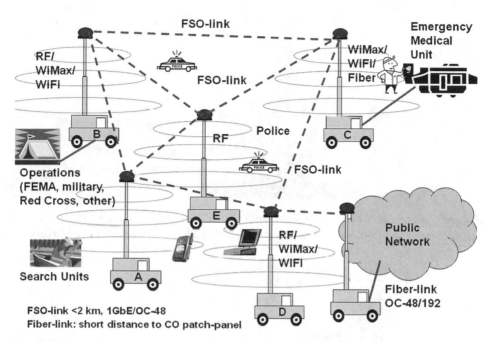

Figure 9.2. Pre-engineered FSO nodes on wheels can quickly construct an ad-hoc mesh FSO communications network in a disaster area to provide broadband access services as well as connectivity with the public network.

Currently, there is no world standard for VLC; the IEEE 802.15 Task group 7 (TG7) is chartered to define the PHY and the MAC layers, and a specialized standard for a visible light ID system has been defined by JEITA (Japan Electronics and Information Technology Industries Association).

VLC possible applications are targeted for short-range (few meters to hundred or so meters) indoor and outdoor LANs that can potentially provide communication, tracking, navigation, communication and entertainment services.

For example, a VLC system could be incorporated with the lighting system:

- Within a building to provide guiding, surveillance, and communications services.
- In streets and highways to provide navigation, tracking, collision avoidance, traffic congestion information, emergency communication and distress call services, and more.

The data rate targeted for VLC application is up to 50 Mb/s, which approaches the limit of standard LEDs. However, *resonant cavity light emitting diodes* (RCLED) operate up to few hundreds of Mb/s and are more efficient than standard LEDs. RCLEDs are vertical structured diodes that consist of a single resonant active layer

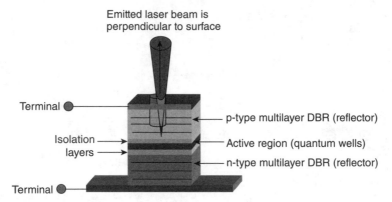

Figure 9.3. A laser beam emanates perpendicular to the surface from a VCSEL. The active region consists of a stack of quantum wells. A RCLED has an active resonant cavity that consists of a single layer.

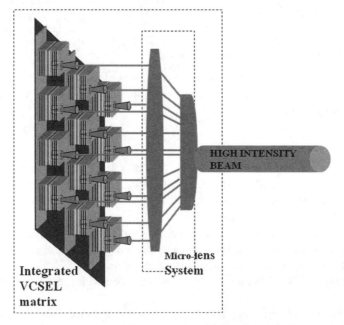

Figure 9.4. Integrating a cluster of LEDs on the same substrate and using a microlens system to produce a high intensity beam.

sandwiched between two Bragg layers; that is, an upright DFB laser, which is also similar to a VCSEL, which has an active layer that consists of a stack of quantum well layers, Figure 9.3.

Now, in fiber-less optical communications (FSO and VLC), although the optical power from each LED may be considered low, a cluster of LEDs may be integrated on the same substrate and it may be combined with a micro-lens system to produce a high intensity optical beam, which is better suited to VLC and FSO applications, Figure 9.4.

9.5 CONCLUSION

FSO technology can be quickly deployed in various topology configurations. Additionally, FSO will benefit from the miniaturization of photonic, optical, and opto-electronic components, as well as by component integration, so that small footprint, longer range, self-tracking and lightweight nodes can establish an ultra high bandwidth reliable network supporting cost-effectively multi-play services (voice, data, Internet, picture and video). When FSO is backed up by RF, then the network is expected to meet uninterrupted service in a plethora of applications, such as residential, enterprise, emergency, permanent and temporary. In addition, FSO is applicable to more specialized communication networks, such as stationary to mobile, mobile to mobile, and also to inter-satellite and deep space.

REFERENCES

1. S.V. Kartalopoulos, *DWDM: Networks, Devices and Technology*, Wiley/IEEE Press, 2003.
2. S.V. Kartalopoulos, *Next Generation SONET/SDH: Voice and Data*, Wiley/IEEE Press, 2004.
3. S.V. Kartalopoulos, *"Free Space Optical Mesh Networks For Broadband Inner-city Communications"*, 10th European Conference on Networks and Optical Communications, NOC 2005, University College London, pp. 344–351, July 5–7, 2005.
4. "Facts About WiMAX And Why Is It The Future of Wireless Broadband". http://www.techpluto.com/wimax-in-detail/. Retrieved 23 November, 2010.
5. "WiMax Forum". http://www.wimaxforum.org/. Retrieved 23 November, 2010.
6. "Overcoming the wire-line bottleneck for 3G wireless services". http://supercommnews.com/wireless/features/wireline_wireless_networks_060305/. Retrieved 23 November, 2010.
7. "High speed Microwave". http://www.wimaxforum.org/technology/faq. Retrieved 23 November, 2010.
8. "FCC Pushes WIMax OK for Katrina Victims, Intel supplies the hardware". http://www.mobilemag.com/content/100/102/C4618/. Retrieved 23 Nov, 2010.
9. "At least two more WiMax handsets coming in 2010", EETimes, 2010-01-04. http://www.eetimes.com/news/latest/showArticle.jhtml?articleID=224201135. Retrieved 23 November, 2010.
10. "IEEE 802.16e Task Group (Mobile WirelessMAN)", http://www.ieee802.org/16/tge/. Retrieved 23 November, 2010.
11. "IEEE 802.16 Task Group d", http://www.ieee802.org/16/tgd/. Retrieved 23 November, 2010.

12. "Wi-Fi Alliance: Organization". http://www.wi-fi.org/organization.php, Retrieved 23 November, 2010.

13. "Wi-Fi Alliance: White Papers" for certification at www.wi-fi.org. http://www.wi-fi.org/wp/ wifi-alliance-certification/. Retrieved 23 November, 2010.

14. "Securing Wi-Fi Wireless Networks with Today's Technologies". http://www.wi-fi.org/files/ wp_4_Securing%20Wireless%20Networks_2-6-03.pdf, Retrieved 23 November, 2010.

ACRONYMS

1GbE = 1 Gbps Ethernet
10GbE = 10 Gbps Ethernet
40GbE = 40 Gbps Ethernet
10GbE = 10 Gbps Ethernet
3G = Third generation
4G = Fourth generation
3R = Re-amplification, Reshaping, and Retiming

AAL = ATM Adaptation Layer
ABR = Available Bit Rate
ACGIH = American Conference of Governmental Industrial Hygienists
AcO = active optics
ACTS = Advanced Communications Technology and Services
ADM = Add-Drop Multiplexer
AdO = adaptive optics
ADSL = Asymmetric Digital Subscriber Line
ADU = application data units
AEL = accessible emission limits
AES = Extensible Authentication Protocol
AIS = alarm indication signal
AIS-L = alarm indication signal for line
AIS-P = alarm indication signal for path
AlGaAs = Aluminum Gallium Arsenide
AMI = Alternate Mark Inversion
AN = Access Node
ANSI = American National Standards Institute
AO = adaptive optics; active optics
AON = All Optical Network

Free Space Optical Networks for Ultra-Broad Band Services, First Edition. Stamatios V. Kartalopoulos.
© 2011 Institute of Electrical and Electronics Engineers. Published 2011 by John Wiley & Sons, Inc.

AP = access point
APD = avalanche photodiode
APDU = Application Protocol Data Unit
APS = Automatic Protection Switching
APSD = automatic power laser shutdown
APT = Acquisition, Pointing, and Tracking
ASE = Amplified Spontaneous Emission
ASIC = application-specific integrated circuit
ASK = Amplitude Shift Keying
ATM = asynchronous transfer mode
AU = Administrative Unit
AUG = Administrative Unit Group
AWG = Array Waveguide Grating

BB = Broadband
BEI = biological exposure indices
BER = bit error rate
BISDN = Broadband Integrated Services Digital Network
BPP = beam parameter product
Bps = Bits per second
BRI = Basic Rate Interface
BS = base station
BT = Burst tolerance
BWIS = beam wander induced scintillation

CAC = connection admission control
CAIDA = cooperative association for Internet data analysis group
CB = coaxial cable
CBR = Constant Bit Rate
CC = Contiguous concatenation
CCMP = Counter Mode with Cipher Block Chaining Message Authentication Code
Protocol
cd = candela or candle
CD = cell delay; Chromatic Dispersion
CDF = client data frame
CDMA = Code Division Multiple Access
CDRH = Center for Devices and Radiological Health
CDV = cell delay variation
CENELEC = European Committee for Electrotechnical Standardization
cHEC = core header error control
CLP = cell loss priority
CM = Communications Module
CMF = client management frames
CMI = Coded Mark Inversion
CMIP = Common Management Information Protocol

CO = Central Office
CP = Customer Premises
CPE = Customer Premises Equipment
CRC = cyclic redundancy check code
CSMA/CA = carrier sense multiple access with collision avoidance
CSMA/CD = carrier sense multiple-access/collision detection
CSP = critical security parameters
CW = continuous wave
CWDM = coarse wavelength division multiplexing

dB = Decibel
dBm = Decibel with 1 mWatt reference level
DBR = distributed Bragg reflectors
DM = deformable mirror
DFB = distributed feedback laser amplifiers
DIRSIG = Digital Imaging and Remote Sensing Image Generatio
DoC = degree of coherence
DOS = denial of service
DRI = Dual Ring Interface
DS3 = Digital service level-3
DSAP = Destination Service Access Point
DSL = digital subscriber lines
DTE = Data Terminal Equipment
DWDM = dense wavelength division multiplexing

E = Extraordinary ray
E1 = A wideband digital facility at 2.048 Mbps, aka CEPT-1
E3 = European digital service level-3
E3 = A broadband digital facility at 34.368 Mbps, aka CEPT-3
E4 = A broadband digital facility at 139.264 Mbps, aka CEPT-4
EAP = Extensible Authentication Protocol
EDC = error detecting code
EDFA = erbium doped fiber amplifier
EFP = environmental failure protection
EFT = environmental failure testing
EIRP = effective isotropic radiated power
EMI = Electromagnetic interference
ESCON = Enterprise Systems Connection
eV = electron Volt

FBG = Fiber Bragg Grating
FC = fiber channel
FCAPS = Fault, Configuration, Accounting, Performance, and Security
FCC = federal communications commission
FDA = US Food and Drug Administration

FDM = Frequency Division Multiplexing
FDMA = Frequency Division Multiple Access
FEC = forward error correction; forward equivalent classes
FICON = fiber connectivity
FOTS = Fiber Optic Transmission System
F-P = Fabry-Perot
FSK = frequency shift keying
FSO = Free Space Optical
FTTH = Fiber to the Home
FWHM = full width at half-maximum

GaAs = Galium Arsenide
GbE = Gigabit Ethernet
Gbps = Gigabits per second = 1000 Mbps
Ge-APD = Germanium APD
GFC = generic flow control
GFP = Generic Framing Procedure
GFP-F = Frame-Mapped GFP
GFP-T = Transparent-Mapped GFP
GMII = Gigabit media independent interface
GMPLS = generic multi-protocol label switching
GPS = Global Positioning Systems
GS = geostationary
GSA = geometrical spreading attenuation
GSL = geometrical spreading loss

HEC = header error control
HeNe = Helium Neon laser
HO = high-order

IDSL = ISDN DSL
IEC = International Electrotechnical Commission
IEEE = Institute of Electrical and Electronics Engineers
IETF = Internet Engineering Task Force
InGaAs = Indium Gallium Arsenide
InGaAsP = Indium Gallium Arsenide Phosporus
INL = inter-node links
INS = Inertial Navigation System
InP = Indium Phosporus
ION = Intelligent optical networks
IP = Internet protocol
IPng = Internet Protocol next generation
IPPM = Internet Protocol Performance Metrics
IPv4 = IP version 4
IPv6 = IP version 6

IR = infrared
ISDN = Integrated Services Digital Network
ISI = intersymbol interference
ISL = inter-satellite links
ITU = International Telecommunications Union

JAXA = Japan aerospace exploration agency
JEITA = Japan Electronics and Information Technology Industries Association

Kbps = Kilobits per second = 1000 bps

LAN = local area network
LAPS = Link Access Procedure SDH SAN = storage area networks
LASER = light amplification by stimulated emission of radiation
LB = Loop Back
LC = Liquid crystal
LCAS = Link Capacity Adjustment Scheme
LDP = label distribution protocol
LED = Light emitting diode
LEOS = low earth orbit satellite systems
LER = label edge router
LGS = laser guide star
LIA = Laser Institute of America
LIDAR = Laser Detection And Ranging nm = nanometer
LL = link loss
LO = lower-order
LOF = loss of frame
LOP = loss of pointer
LoS = line-of sight
LOS = loss of signal
LSR = label-switched router
LTE = Long term evolution

MAC = media access control
MAN = metropolitan area network
MBE = molecular beam epitaxy
Mbps = Megabits per second (1000 Kbps)
MBS = Maximum burst size
MDI = Medium dependent interface sublayer
MEMS = micro-electro-mechanical systems
Mhz = Megaherzt (10^6 Hertz)
MIMO = Multiple Input–Multiple Output
MMF = multi-mode fiber
MOCVD = metal organic chemical vapor deposition
MPE = maximum permissible exposure

MPLS = multi-protocol label switching
MPλS = Multiprotocol lambda (wavelength) switching
MQW = multiple quantum well
MQWL = multiple quantum well lasers
MS = mobile station
msec = milliseconds
μsec = microseconds
MSPP = multiservice provisioning platform
MSSP = multiservice switching platform
MUX = Multiplexer
mW = milliWatts

NE = network elements
NGI = Next Generation Internet
NG-OR = Next Generation Optical Ring
NG-S = next generation SDH
NIC = network interface circuit
NNI = network to network interface
nm = nanometer
NRM = network reference model
nrt-VBR = Non-real time Variable Bit Rate
NRZ = Non Return to Zero
ns = nanosecond

O = Ordinary ray
O_3 = Ozone
OA = Optical Amplifier
OA&M = operations, administration, and maintenance
OADM = optical add/drop multiplexer
OC = Optical Carrier
OC3 = optical carrier -3
OC-N = optical carrier-N (N = 1, 3, 12, 48, 192)
ODL = Optical Data Link
O-E = Optical to Electrical conversion
O-E-O = optical-to electrical to optical
OFA = optical fiber amplifiers
OFDMA = Orthogonal Frequency-Division Multiple Access
OFS = Optical Fiber System
OH = Overhead
OLS = Optical Line System
ONSR = optical signal to noise ratio
O-O = optical to optical
OOK = on-off keying
OSI = open system interconnect

OSI-RM = Open System Interconnect Reference Model
OSNR = Optical Signal-to-Noise Ratio

PC = personal computers
PCR = Peak cell rate
PCS = physical coding sublayer
PD = Photodiode; Propagation Delay
PDF = probability density function
PDU = Protocol Data Unit
PHY = Physical Layer
PIN = positive intrinsic negative
PLS = physical layer signaling
PM = performance monitoring
PMA = physical medium attachment
PMD = physical medium dependent
PMI = Physical Medium independent interface
POH = Path Overhead
PON = Passive Optical Network
PPMN = path protected mesh network
PPMV = Parts per million by volume
PSDN = public switched digital network
PSK = phase-shift keying
PSP = Principal States of Polarization
PTE = path terminating network elements
PTI = payload type indication
PtP = point-to-point

QoS = quality of service
QWL = quantum well lasers

RCLED = resonant cavity LED
RDI-L = received defect indication for line
RDI-P = received defect indication for path
RF = radio frequency
RSL = receiver signal level
RTCP = Real Time Control Protocol
RTP = Real Time Transport
RTT = round trip time
rt-VBR = Real time Variable Bit Rate
RVP = runway visual range

SA_T = surface area of the transmit aperture
SA_R = surface area of the receive aperture
SAPI = service access point identifier

SBR = solar background radiation
SCR = Sustainable cell rate
SDH = Synchronous digital hierarchy
SEL = surface emitting lasers
Si-PIN = Silicon APD
SLA = service level agreement
SL-n = Security Level n
SMF = single-mode fiber
SMSR = single mode suppression ratio
SNR = Signal to Noise Ratio
SOA = semiconductor optical amplifiers
SoF = Start of frame
SOFDMA = Scalable OFDMA
SONET = Synchronous optical network
SoPSK = state of polarization shift keying
SPE = synchronous payload envelope
ST = shielded twisted pair
STM-1 = synchronous transport module level-1
STM-M = synchronous transport module level-M (M-1, 4, 16, 64)
STP = standard temperature and pressure
STS-N = synchronous transport signal level-N

T1 = A digital carrier facility used to transmit a DS1 signal at 1.544 Mbps
T3 = A digital carrier facility used to transmit a DS3 signal at 45 Mbps
Tbps = Terabits per second = 1000 Gbps
TCP = Transmission Control Protocol
TCP/IP = Transmission Control Protocol/Internet Protocol
TDM = Time Division Multiplexing
TEC = thermoelectric cooler
TEM_{00} = fundamental transverse electrical mode
TEM_{mn} = transverse electrical mode mn
TG7 = IEEE 802.15 Task group 7
Thz = Tera-hertz (1000 Ghz)
TIA = transimpedance amplifiers
TKIP = Temporal Key Integrity Protocol
TLV = threshold limit values
TM = transport module
TOH = Transport OverHead
TP = Twisted Pair
TU = tributary unit
TU-n = Tributary Unit level n; n = 11, 12, 2, or 3
TUG-n = Tributary Unit Group n; n = 2 or 3

UBR = Undefined Bit Rate
UDP = User Datagram Protocol

UHF = ultra-high frequency UV = ultraviolet
UL/DL = Up Link/Down Link
UMTS = Universal Mobile Telecommunications System
UNI = user to network interface
UPI = user payload type indication
UTP = unshielded twisted pair cable
UV = Ultra-violet

VBR = Variable Bit Rate
VC = virtual channel
VCC = Virtual Channel Connection
VCI = Virtual channel Identifier
VC-n = Virtual Container level n (n = 2, 3, 4, 11, or 12)
VCSEL = vertical cavity surface emitting lasers
VDSL = Very-high-bit rate DSL
VHF = very-high frequency
Video-o-IP = Video over the Internet protocol
VLC = visible light communications
VoIP = Voice over the Internet protocol
VP = virtual path
VPC = Virtual Path Connection
VPI = virtual path Identifier
VT = virtual tributary

WADM = Wavelength Add-Drop Multiplexer
WAN = wide area networks
WDM = Wavelength division multiplexing
Wi-Fi = "Wireless Fidelity"; a protocol defined by the Wi-Fi Alliance
WLAN = Wireless LAN
WiMAX = Worldwide Interoperability for Microwave Access
WPA = Wi-Fi Protected Access
WtW = window-to-window
WvoIP = Wireless Voice over the Internet

XOR = exclusive OR

INDEX

Free Space Optical Networks for Ultra-Broad Band Services, First Edition. Stamatios V. Kartalopoulos.
© 2011 Institute of Electrical and Electronics Engineers. Published 2011 by John Wiley & Sons, Inc.